Schools Council
Educational Use of Living Organisms
General Editor: P J Kelly

Animal Accommodation for Schools

Author: J D Wray

Contributor: J F Gaitens

The English Universities Press Ltd

Schools Council Educational Use of Living Organisms Project

Director: P.J. Kelly

Research Fellow: J.D. Wray

This project was established at the Centre for Science Education, Chelsea College, in 1969. Its main aims have been to determine the needs of schools with respect to living organisms, to evaluate the usefulness of various kinds of organisms for educational purposes, and to devise maintenance techniques and teaching procedures for the effective use of appropriate species.

ISBN 0 340 17049 2

First printed 1974

The English Universities Press Ltd
St Paul's House, Warwick Lane, London EC4P 4AH

Printed and bound in Great Britain by Chigwell Press Ltd, Albert Road, Buckhurst Hill, Essex.

Computer Typesetting by Print Origination, Bootle, Lancs, L20 6NS

Contents

Preface

The use of living organisms in schools, while not new, has received considerable emphasis in recent developments in the teaching of biology, environmental studies and allied subjects in colleges and secondary schools, and in many aspects of primary school work. It is a change reflecting the much wider movement in educational thinking which acknowledges both the interest and delight that young people can gain from animals and plants and the importance of fostering an appreciation of the scientific, social, aesthetic and moral issues involved in the study of life and the natural environment. It is a change, also, that presents some very real – but not insurmountable – practical problems for schools.

The implications of using living organisms for education are basically threefold. There needs to be an adequate supply of appropriate organisms and they should be kept healthy and be able to live as naturally as possible. Accommodation for the organisms should allow them to be observed and studied easily but with respect. Maintenance of the organisms should not place an undue burden on teachers and technicians. The Educational Use of Living Organisms publications are intended to assist teachers (including, of course, student teachers), technicians and administrators to control with these implications.

The books in the series deal with the principles which underlie the effective educational use of living things. They also provide information to help teachers integrate work with organisms into their courses and to cope with the practical, day-to-day problems involved. For teachers and administrators there are technical details of value for planning facilities, and annotated bibliographies provide the guidelines for more detailed studies if they are required. In addition, posters and slide transparencies for use with pupils have been produced complementary to the books.

The series has been produced as part of the work of the Educational Use of Living Organisms project. The project was initiated by the Institute of Biology and Royal Society Biological Education Committee and was established as a Schools Council Project at the Centre for Science Education, Chelsea College (University of London), in 1969. While the project's major financial support was from the Schools Council, the Nuffield Foundation and Harris Biological Supplies have also given generous contributions.

Many people have assisted in a personal capacity and we have particular regard to the interest shown by Mr D J B Copp, Mr J A Barker, Mr T A Gerrard, Mr J F Haller, Mr O J E Pullen and Dr C H Selby. by Mr J A Barker, Mr D J B Copp, Mr T A Gerrard, Mr J D Wray undertook much of the research of the project and provided much of the information on which the books are based. In this, he was admirably supported by Mr J B Green as technician and Miss M Hoy and later Mrs E Barton as the project's secretary.

The authors of the books very kindly have been most tolerant of their editor. I would like to express my gratitude to them and our publisher for both their help and forbearance.

P J Kelly
Director, Educational Use of Living Organisms Project.

Note.
The accommodation described here is only for the housing and husbandry of animals and for their use in instruction which does not require the issue of a licence from the Home Office to comply with the 1876 Cruelty to Animals Act.

Chapter 1 Guidelines and standards

Using animals in a school involves two seemingly opposed requirements. On the one hand pupils should be able to observe and handle the organisms. On the other, the animals must be kept healthy and this means ensuring that they are unaffected by wide fluctuations in environmental conditions, which are likely to occur in classrooms and laboratories, that the chances of transmitting disease are minimal, and that the animals are not caused undue stress by the pupils. The maintenance and use of the animals needs to be carefully regulated and for this special accommodation is required.

The requirements for schools will of course vary, depending on the facilities available and the needs to be served. For this reason the aim of this book is to provide information and suggestions which will provide guidelines for the design and use of accommodation suitable for a wide range of circumstances.

Considerations for planning

Whatever accommodation is being designed some fifteen points need to be considered. They are listed here and can be used as a check list to ensure that they are taken into account in the planning.

1 *Purpose of the accommodation:* This should be well defined. All else depends on it.
2 *Range of species to be maintained and likely numbers:* This will indicate the size of accommodation that is needed and the facilities required in it.
3 *Staff and pupil access and use:* Who will have access? To what extent is it intended that pupils should look after the organisms? What supervision will be given?
4 *Finance:* Estimate the various costs and find out what money is available. Some compromise will be inevitable!
5 *Site:* What sort of space is available? Calculate the space required allowing for easy maintenance of the structure. Consider the possible effect of external conditions on the environment in the accommodation. Factors such as temperature and light will depend on the aspect, position and exposure of the accommodation.
6 *Relationship to classrooms, laboratories, greenhouses and gardens:* Consider the access to water, electricity and other services, the convenience of transporting stock to and from laboratories, the problem of security, possible noise disturbance and possible contamination from pesticides and other chemicals used in the laboratories, greenhouses and gardens.
7 *Approach and entrance:* Entrances should be arranged to facilitate the use of trolleys etc., and to be in such a position as to provide maximum security.
8 *Structure:* Assess the advantages and disadvantages both short and long term of a site-built or converted structure with that of a prefabricated structure. Financial considerations may be crucial here. Ease of cleaning and servicing are very important.
9 *Layout:* Consider possible arrangements of the stock, breeding, working and observation areas, paying attention to ease of cleaning.
10 *Services:* Estimate requirements with respect to water, electricity and drainage. The position of the accommodation may have a considerable influence. Allow for the maintenance of services.
11 *Macro-environmental control:* Investigate the means of controlling temperature, ventilation, noise, light and humidity within the accommodation.
12 *Equipment:* Consider racking or shelving, the provision of cages, protective clothing and cleaning materials.
13 *Storage:* Allow for the safe storage of food, litter, nesting materials and spare cages.
14 *Security:* Consider the regulation of access to the accommodation, fire precautions and the prevention of injury to animals and humans.
15 *Management and Use:* Consider the health of staff, pupils and animals, the control of possible infection and the means of effective cleaning and

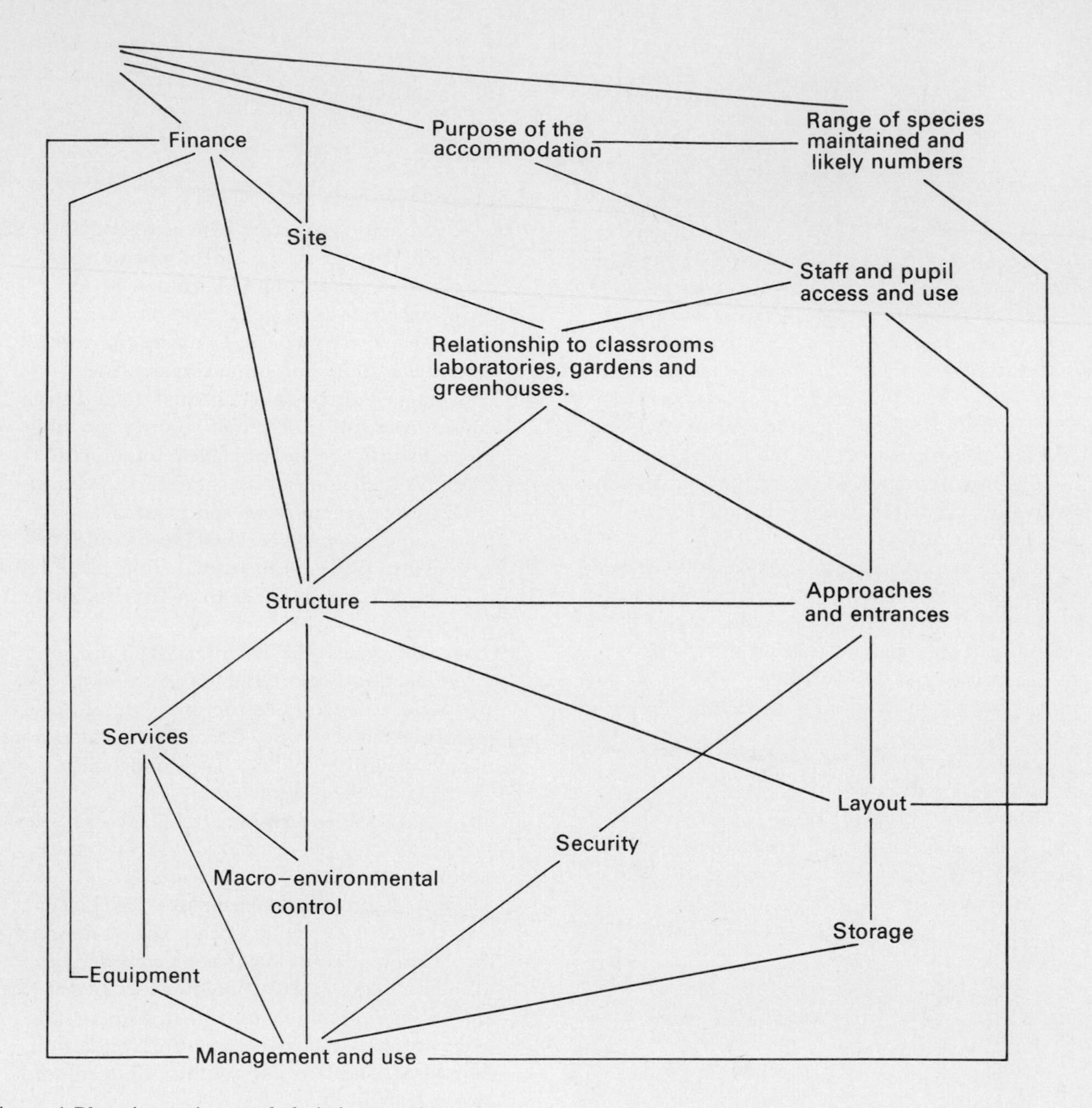

Figure 1 Planning points and their interaction.

maintenance. Decide whether or not pupils are to be allowed to handle and observe the organisms in the accommodation and whether organisms are to be taken into laboratories and/or classrooms.

In dealing with each of these planning points others will have to be taken into account. The association between the planning points are indicated in Fig 1.

Suitable and unsuitable accommodation
Whether or not accommodation can be judged to be suitable depends on many points of detail. These will be dealt with in due course but, in the early stages of planning the accommodation, it is useful to have an idea of the overall standards that need to be borne in mind. Figures 2-8 show accommodation of widely different types, indicating some of the things required and some that ought to be avoided.

Figure 2 Detached brick-built house.
A well constructed brick-built house. Note the metal protection of the door base and the ventilator well placed near the roof. Type A1.

Figure 3 Wooden hut.
A small wooden hut 3.6 x 2.4m., which has been well adapted. It has electric heating, good ventilation and is adequately insulated by the use of insulation board. An unusual feature for small wooden houses is the provision of washing facilities. Unfortunately it is a considerable distance from the biology department. Type B3.

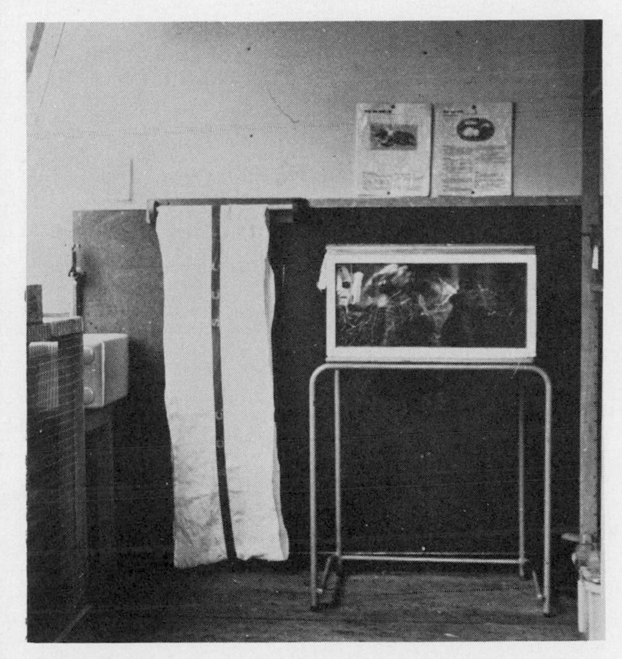

Figure 4 Interior of the wooden hut shown in Fig. 3.

Figure 5 Integrated house.
A small house as one end of a lean-to greenhouse, not in itself a good feature. It has a glass roof and large windows none of which are double glazed and temperature control is a problem. It is often too hot in the summer and occasionally too cold in the winter despite electric heating: A rodent barrier should be fixed to the door frame. Type A2.

Figure 6 Integrated house.
A large house originally intended as a greenhouse. Ample size but with problems of temperature control. Some space is lost since the animals cannot be placed near the windows particularly in the winter. Type A2.

Figure 7 Concrete garage.
Use of a concrete garage that is completely uninsulated. Maintaining a suitable temperature in the winter is difficult and very expensive. The large gap under the main doors is difficult to seal. Type B2.

Figure 8 Converted room.
A conversion of a room in a biology department. Adequately heated and ventilated, but far too small and with the top shelves difficult to reach. Type C. Situated in a position similar to the accommodation in Figure 26.

What animals might be kept?
For normal school use a collection of animals including small mammals, e.g. mice, rats, Syrian hamsters, Mongolian gerbils, guinea pigs and rabbits, a few small birds e.g. zebra finches, a few fish, reptiles and amphibia, some smaller animals including the larger insects and possibly protozoa would be suitable.[76]

All of these can be kept in the animal house although animals maintained in self-heated containers such as vivaria, aquaria and insect cages may be most conveniently kept in the laboratory providing there is space, little danger from fluctuations in temperature and the health hazards to staff and pupils are not great.

Small insects e.g. *Tribolium, Drosophila* and ants, and other arthropods e.g. woodlice, should not be kept in the animal house unless special precautions are taken to prevent their escape or to restrict their movement should it occur.

Housing a variety of species together is, of course, not ideal. In schools it has to be accepted, however, so it is necessary to make special provisions. Inevitably a compromise must be made between what is desirable and what is practically possible (Fig 9).

Figure 9 Housing a variety of species.
The housing of a variety of species on commercially available wooden shelving and metal stand. Note the wide range of containers, the uppermost being of an unsuitable painted metal type.

Chapter 2 Finance

Money will inevitably be a major consideration when contemplating building animal accommodation. For each animal house the cost will vary, depending on the type of material that is used, the needs it is expected to serve and a variety of other factors. Nevertheless, before designing the detail of an animal house it is important to have some idea of the relative initial costs of different types of structure, and of the proportions of costs for construction, installation of services, equipment, purchase of stocks and subsequent recurrent costs.

Type of structure:
There is considerable variation in the cost of purpose-built animal accommodation for schools depending mainly on the type of structure employed. Broadly speaking, there are three possible types of structure.

Type A: Specially designed buildings entirely constructed on the site. These may be:
A1: detached from the school buildings, or
A2: integrated with them and having an entrance from within the school.

Type B: Prefabricated structures assembled on the site. These may be:
B1: constructed of glass-fibre,
B2: constructed essentially of precast concrete blocks (e.g. concrete 'garages'),
B3: constructed essentially of wood (wooden huts and sheds)

Type C: Special conversions of suitable spaces within the school building.

Approximate costs per square metre of floor area for each of these types, excluding the cost of equipment and stock, are as follows. (The prices are for 1971. They will, no doubt, have increased but it is unlikely that they will change relative to one another.)

Type A: Site constructed

A1	Detached	£33 – £110+
A2	Integrated	£33 – £88+

Type B: Prefabricated, (excluding the cost of installation of services and possible delivery and erection charges).

B1	Glass-fibre	£36 – £45
B2	Pre-cast concrete block construction	£9 – £17
B3	Wooden	£11 – £20

Type C: Conversions.
These are very variable depending on the degree of conversion required £22 – £66+

The initial cost can be broken down in the following proportions:

construction	40 – 60%
installation of services	20 – 40%
equipment	15 – 25%
purchase of stocks	5%

In addition, recurrent costs of approximately 5% will have to be contemplated.

Assessing costs in detail
It is important when doing this not to inadvertently leave out items and it is helpful to follow a standard list of items such as those used in making bills of quantities or briefs for architects.[5] Of course, not all of the items may apply to each individual case.

Suitable brief lists for estimating the costs of the structure would be:

Type A: Site constructed
Preliminaries: the introduction to the list of items showing quantity and type of work in a building contract (some items can be priced at the outset).
Insurance
Contingencies: sums reserved for unforseen expenditure used or deducted in whole or in part
Work below floor finish: foundations
Structural elements: load bearing frame, roof, external walls, windows (if any), external doors, internal walls and partitions, internal doors
Ironmongery
Finishes and fittings: wall, floor and ceiling finishes,

fittings to include all built-in furniture

Type B: Prefabricated
Purchase cost
Delivery charges if any
Foundations and base where necessary
Erection charges if any
Insulation, waterproofing and finishing off inside as necessary

Type C: Conversions
Preliminaries
Insurance
Contingencies
Work below floor level if necessary
Alteration of structural elements if necessary
Fittings
Making good the finishes

In estimating the cost of installing equipment and services the following items need to be considered. Note that even with converted accommodation, services may have to be installed or altered.

Services
Water
- piping
- taps
- hose connections
- sinks and/or dunk tanks

Electricity
- cabling
- suitable fusing devices
- waterproof switches both on/off and timed.
- heating units, lighting units, ventilation units } with suitable waterproof connections.
- additional waterproof outlets for small pieces of equipment.

Drainage
- gridding
- removable traps
- piping

Equipment
Rackings and/or shelving and supports
Work surface at suitable height
Cages including water bottles and food hoppers
Vivaria
Aquaria
Rodent barriers
Trolleys
Storage bins
Brushes, cleaning equipment and material
Dispensers for liquid soaps or disinfectants
Hosepipes
Protective clothing
Fire-extinguishers
Incinerator
First-aid materials for human and other animals.
Equipment necessary for the maintenance of certain animals e.g. aquarium heaters and air pumps for fish and nail clippers for small mammals.

The cost of purchase and delivery of the animals will vary depending on the purpose and use of the accommodation. The catalogues and lists of appropriate suppliers (see Appendix 1) should be consulted.

Expenditure after the installation of the accommodation will consist of the recurrent costs of heating and lighting, of cleaning and disinfecting the building and of maintaining the building, the stock, the equipment, and the services. The cost of maintaining the stock will include expenditure on food, nesting and litter materials, veterinary services, first aid materials, disinfectants and replacement stock at regular intervals. Note that prefabricated structures, particularly wooden ones, are likely to be more expensive to maintain than site-constructed buildings or conversions.

Further details of cost and planning of buildings may be found in reference 5 and information about the cost factors to be considered when designing research or commercial animal houses, some of which are relevant to accommodation in schools, may be found in references 6, 11 and 14.

Chapter 3 Situation, structure, layout and services

Situation

The ideal site for an animal house would be sheltered from the wind and direct exposure to the sun. It would be on the same level as, and close to the laboratories and classrooms in which the organisms may be used, yet sufficiently isolated to avoid undue disturbance to the animals by noise and people.

If the accommodation is detached, consideration must be given to the problems and convenience of transporting animals, cages and food materials etc., to the cost of installing and connecting essential services, to the security of the accommodation, to the maintenance of structure and services, and to blending the appearance of the animal house with that of the school buildings.

It is undesirable to have an animal house and a greenhouse too close together. Environmental requirements inside and outside such structures are widely different and, contamination of the animal accommodation by pesticides and composts used in the greenhouse can prove fatal to animals.

Structure

The design of an animal house, like that of any addition to school buildings, will be subject to local authority and other building regulations. The information that follows deals essentially with the structure and internal design of the accommodation, not with its external appearance. It can be used to prepare a brief for an architect who will then be in a position to design the animal house within the limits imposed by regulations and other external considerations.

Site-constructed accommodation and conversions

Walls

Internal walls are best constructed of fair faced brick (11 cm) or precast blocks (10 cm), which should be strong enough to carry brackets for shelving, racking and heaters. A covering of hard plaster or cement should be applied and a reliable washable finish used. The join between the wall and floor and internal angles should be coved to avoid a dust trap.

The wall area near sinks may be tiled, but the jointing of the wall tiles may not remain impervious to water. If cages are to be hosed down then it is best to cover this area with asphalt, welded PVC sheet or other applied plastic materials.

Exposed angles of the wall which may be subject to damage from the impact of trolleys should be protected, possibly by the use of non-ferrous metal strips. Special thermal insulation of the walls will not be necessary unless the site is very exposed.

The avoidance of cracks and crevices in which dirt and small insects can lodge is very important in construction and surface finish.

Roofing and ceiling

Insulation of the roof is of the greatest importance. A maximum thermal transmittance or 'U' value for this type of accommodation of 1.135 $W/m^2\ ^\circ C$ (= 0.2 $Btu/ft^2\ ^\circ F$) is recommended[24], and values lower than this can be achieved by careful insulation (see Appendix 2). Unless the roof is part of the existing school structure the use of pitched roofs of slates or tiles is not justified. The join between ceiling and wall should be coved. The ceiling should be plastered and finished with washable emulsion paint.

Windows

For the animals likely to be housed, natural daylight is not necessary. It is possible to have a windowless building. However it is useful to have at least one window which can be opened since it may have to be used as a means of ventilation during maintenance or failure of ventilator fans.

Windows, except those specially provided for observation purposes, should be situated at least 2 m (6.5 ft) above the floor level so that they do not reduce the effective use of the walls for shelving and racking (Fig 10). They should be metal-framed and, if possible, double glazed. Those that are to be opened must have a mesh screen fitted to the fixed frame to make them insect- and rodent-proof when open. Translucent panes made of glass bricks can be used for partial illumination.

Doors
It is desirable to have only one door to limit draughts, disturbance, and possible transmission of infection.

The door frame should be of metal and accord to British Standards (BS 1245-1951). The door can be made of metal or wood and should open outwards. If a wooden frame or door is used it should be protected by metal, preferably stainless steel or galvanised iron, to a height of 1 m so that it cannot be gnawed by rodents. Draughts from the door should be reduced by fitting a metal draught excluder strip. If a window is needed in the door it should be of wired glass as a safety precaution. A rodent barrier (Fig 11) is necessary on the inside of the door frame. This is a removable, rigid, galvanised iron plate about 0.5 m high fitting into metal guide slots attached to the frame.

Floor
This must be impervious to and unaffected by water, urine, disinfectants and detergents. It must also be resistant to abrasion by footwear and movable equipment and not be slippery when wet. A variety of materials may be used for the construction and finish of the floor.[1] The cheapest satisfactory material is smooth cement finished with a sealing compound containing sharp sand or carborundum. Other possible surfaces are asphalt, 'Terrazzo', new plastic compounds and floor finishes based on polymer compounds such as polyvinyl chloride and acetate.[2] The use of either quarry tiles or plastic tiles is not to be recommended since the joints create areas which may become filled with micro-organisms. A shallow depression to accommodate a thin mat, may be let into the floor just inside the doorway. When in use this mat may be kept moistened with a suitable disinfectant solution. Alternatively a scraper mat can be used.

Information on the gradients necessary for the drainage of the floor is given in the section on services.

Prefabricated structures
Prefabricated structures constructed entirely from a sandwich of glass fibre and insulating material are very suitable in that they can be easily cleaned, are totally waterproof and effectively insulated. However, those available at present are relatively expensive and require modification to allow simple equipment to be used for adequate ventilation and heating.

Figure 10 Windows.
Metal framed double glazed window fitted just below the roof allowing full use of the wall for shelving or racking.

Figure 11 Rodent barrier.
A metal rodent barrier fitted to the door frame. Note that the base of the frame is protected by metal.

Figure 12 Supports for a small wooden animal house. Wooden animal house with wooden floors must be raised on brick or concrete piles, containing a damp proof course, to a height which will prevent rodents from nesting underneath.

Prefabricated concrete or wooden structures are not altogether desirable but some can be modified to make acceptable accommodation. Prefabricated concrete garages can be obtained with the usual car door or doors replaced with a solid end with little, if any, increase in cost. Thus, if they have a side door of adequate size they can provide suitable basic structures.

Small prefabricated wooden structures are the least satisfactory. They are difficult to clean and to waterproof. They are not recommended but, if used, should be made of the highest quality timbers fully proofed against fungal and insect attack.

Prefabricated concrete or wooden structures should be stood on a special concrete raft which if waterproofed with a damp-proof membrane can provide an adequate floor. Drainage gradients could be incorporated when the raft is laid (see p 15). Gaps between the base of the structure and the foundation must be sealed to prevent the entry of rodents and insects and to be hygienic. Ramps at the door will then be necessary if trolleys are to be used.

Wooden floors in wooden huts are not very suitable but if used the entire structure should be raised, by standing the joists on brick or concrete piles containing a damp proof course, at least 20 cm above ground level. This will allow access underneath and prevent rodents from nesting (Fig 12).

Thermal insulation is essential in the reduction of temperature fluctuation, economy of heating costs and prevention of condensation.[3; 24] To facilitate cleaning a waterproof surface to such insulation is necessary.

Covering the inside of the walls and roof with insulating building panels is the most effective means of giving a long lasting insulation layer which will withstand the conditions without deterioration. There is a range of such insulating panels available made of rigid polyurethane foam with integral liners and vapour barriers in various thicknesses. Those faced with polythene are most suitable since this face is waterproof and washable (for example, Imperial Chemical Industries Ltd 'Purlboard' P.2). This insulation layer can be bolted, screwed or nailed to the basic structure or to wooden battens attached to the walls of concrete structures. The joints between these boards can be sealed with special self-adhesive tape to complete the vapour seal and waterproofing surface. Rodent barriers should be fitted to the doors as detailed on page 13.

Internal Layout

Minimum internal dimensions of 7 x 4 m (20 x 12 ft) and a height of 3 m (10 ft) are recommended for an animal house if it is to be of real value to a school. The interior design should be simple and flexible enough to allow a sufficient variety of species to be maintained and used effectively.

Particular attention should be paid to the ease of cleaning and servicing. It is in these respects that prefabricated structures, with the possible exception of glass-fibre structures, are deficient.

In the ideal internal layout there would be six separated areas:

1 An acclimatisation area for the reception of new stock or animals previously used in the school laboratories or classrooms.
2 A quarantine area for the confinement of sick animals.
3 A stock and breeding area to hold demonstration and breeding animals.
4 A 'clean' storage area to hold food, litter, nesting materials and clean cages etc.

5 A 'dirty' area in which cage cleaning, waste disposal and similar activities can take place.
6 An area in which pupils can observe the animals without unduly disturbing them or increasing the risks of contamination.
Further information on design and layout will be found in references 6, 7, 11, 39 and 40.

Services
Three basic services must be considered: water, drainage and electricity.

Water
Hot and cold running water should be available. An electric water heater will be necessary if hot water is not available from the school system during weekends and holidays. Piping should be concealed in the walls as far as possible since exposed piping creates a dust trap and is unhygienic. Connection of the water system should comply with the relevant regulations concerning rising mains water.

It is suggested, for reasons of economy, that a small sink for hand washing and a large 'dunk tank' for the disinfection of cages be provided. This tank can be made of plastic, glass-fibre or metal and should be at a convenient working height. The dimensions of this tank should be such that the bases of large cages (ideally the entire cage) can be totally immersed for disinfecting. Suggested minimum dimensions are 1 m x 0.6 m x 0.6 m (3 ft x 2 ft x 2 ft) (Fig 13).

If the cold-water tap over the hand-sink is fitted with a hose connection, the tank can be filled using a hose. Alternatively a specially installed cold-water tap over the tank will be necessary. A removable mesh grid or filter to retain solid waste should be fitted to the tank outlet.

Drainage
Drains should be of a size adequate to cope with the rate of water flow when all the taps are turned on so that the cleaning of floor and walls by hosing with cold water can be done without flooding.

It is essential that the top drainage outlet is below floor level and this should be carefully checked with reference to the anticipated floor level, during the design stage. Drainage channels in the floor may be provided (Fig 14). Gradients on the floor and in the channels should not be less than a 1.0 cm drop in 1 m (i.e. 1 in 100). Drains and floor outlets should have a

Figure 13 Disinfection unit.
Large 'dunk tank' and small sink in one combined unit. Cages can be sterilised in the 'dunk tank'.

Figure 14 Floor drainage.
Drainage channel in concrete floor. The wooden slats normally in the channel have been removed for clarity.

Figure 15 Electric switches.
Electric switching should be waterproof if the walls are to be washed with water. Note the master switch and thermostat in the top photograph.

watertight joint with the floor finish, be fitted with a removable mesh trap of filter to retain solid-waste, and must be secure against the entry of rodents from outside. Adequate, readily accessible, cleaning points should be incorporated in the drains and it is desirable to keep this system separate from other drainage systems to avoid problems resulting from blockage and backflow. Drains made of polyvinyl chloride (PVC) are most suitable.
Further information may be found in reference 8.

Electricity
Electrical power will be required for lighting, heating ventilation and for ancillary equipment, such as air pumps. Details of these requirements are given in Chapter 5. All electrical cabling should be concealed in the walls as far as possible. It is recommended that cabling should not be in conduits since internal condensation can be a problem. Connections will obviously be required for lighting, heating and ventilating apparatus and additional socket outlets should be provided, in the most advantageous positions, preferably above the level where the animals are housed, for small electric pumps, small heaters and movable lights.

The electric switches, controls, socket outlets and apparatus must be waterproof, so that the walls may be effectively washed (Fig 15). Conventional outdoor waterproof switches (lower photograph) and socket outlets can be used or floor socket outlets could be used with the hinged flap made to close on a waterproof gasket. If switches and sockets are surface mounted they are less susceptible to water. Time switches and the controls for extractor fans should be enclosed in a suitable waterproof box. All electrical apparatus must be adequately earthed.

It is advisable to have the electrical circuit to the accommodation entirely separate from that to the rest of the school so that an uninterrupted supply is available. As an additional safety measure a master switch should be incorporated in the circuit (upper photograph) and a suitable earth leakage trip or manual resetting contact breaker may be incorporated.

Teachers wishing to connect electrical apparatus should check with their Local Education Authority that they are allowed to do so. It is essential that all wiring is checked, if not installed, by a qualified electrician. Further information may be found in references 10 and 37.

Chapter 4 Controlling the environment

The overall environment inside an animal house can be termed its macro-environment to distinguish it from the micro-environments within the cages and other units in which different species are kept. Each species will have its own environmental requirements and the macro-environment should be seen as a compromise of these within which a range of micro-environments can be maintained.

Temperature, ventilation, noise, light and humidity are the major factors that contribute to the macro-environment and need to be controlled within prescribed limits.

Temperature

Allowing for the variation in temperature which will occur in the accommodation and for the fact that the micro-environmental temperature within individual cages can be altered by the animals themselves and/or by the use of additional heaters, it is suggested that an optimum temperature for the macro-environment of 20°C (68°F) and a permissible range from 18°C to 24°C (64°F to 75°F) is desirable. The temperature should be measured at a height of 0.5-1.0 m above the floor and away from the entrances.

Electric heating is most satisfactory to maintain this level of temperature. The house can be heated either by heating the incoming air or by the use of space heaters such as tubular heaters. If the house is integral with the school buildings, use may be made of the existing central heating system provided an additional electrical heating system is built in, adequate to maintain the temperature during the night-time, weekends and holidays when the school central heating system may be reduced or switched off. The required capacity of the heating system will depend on a number of factors, such as the exposure of the site, the orientation of the accommodation, the adequacy of insulation and draught proofing, presence or absence of windows, the cubic capacity of the accommodation and the total number and variety of animals housed.

The heating source should have sufficient capacity to raise the room temperature to the desired level from the minimum outside temperature which might be experienced in the district and not from 0°C (32°F) as with most other premises.[24] Tubular electric heaters controlled by thermostats are very satisfactory. They should be fixed to the walls approximately 30 cm (1 ft) from the floor and between 10-15 cm (4-6 inches) out from the wall to facilitate cleaning and repair and positioned to give an even distribution of heat (Fig 16). Wiring, and conduits if used, should be concealed in the walls as far as possible and the wiring connections, the heaters and the thermostats must be waterproof. Adequate earthing of the heaters and thermostats is essential.

Radiant heaters tend to give a very uneven heat distribution and are not recommended except as a special heat source for chicks or quarantine cages. Fan blown heaters should not be used as the main source of heat and bulky convector heaters are not desirable.

Additional heat sources for cages, vivaria and other small containers may be provided if necessary using incandescent lamps, small infra-red heaters of the chick-brooder type, or small heaters of a type used

Figure 16 Electric heating.
Tubular electric heaters in a small wooden animal house protected by a removable metal mesh.

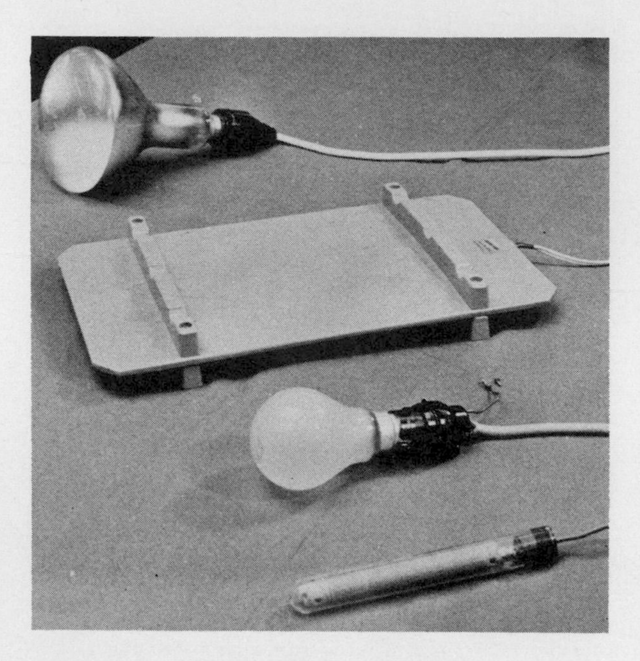

Figure 17 Alternative sources of heat.
An infra-red heater of the chick-brooder type, a small seed tray heater, an incandescent lamp and immersion heater. All these must be adequately earthed when in use.

for heating seed trays (Fig 17). All these must be adequately earthed.

Immersion heaters which are safe even if the glass envelope is broken when submerged, are now commercially available and should be used for heating aquaria. Alternatively, base heaters may be used.

Further information on heating and temperature ranges may be found in references 3, 20, 24, 29, 35, 37, 38, 39 and 40.

Details of the method of estimating the required heating capacity for detached accommodation are given in Appendix 2.

Ventilation

The basic function of ventilation are[24]:

1 To give an optimum and even temperature.
2 To maintain suitable humidity conditions for the animals' comfort, and in respect of some species, to prevent diseases.
3 To dissipate the heat produced by the animals.
4 To remove animal odours.
5 To provide sufficient air changes to remove rapidly any dangerous air-borne organisms which might spread infection among the animals.

It is thought by some authorities that the provision of oxygen for respiration and the removal of carbon-dioxide is *not* one of the main functions of ventilation.

Air movement in the accommodation should be as uniform as possible and draughts are to be avoided at all costs. A linear air speed in the free areas of the accommodation of between 0.10-0.15 metres per second (20-30 ft per minute) at a temperature of 18.5°C (65°F) is suitable, increasing in warm weather with temperatures above 21°C (70°F).

The rate of ventilation required will depend on a number of factors, including the exposure of the site, the seasonal temperature, the efficiency of thermal insulation, and the range of the animal species maintained and their heat production. Thus it is not possible to lay down a standard air change rate[24] for all animal accommodation. However, in general, a minimum of 6 air changes per hour in winter and 12 per hour in the summer will be satisfactory providing that these rates give good thermal control.

For effective ventilation it will generally be necessary to use electric extractor or exhaust fans. The choice of fan unit depends on the cubic capacity of the accommodation and the desired ventilation rate. The fans must be capable of running continuously for very long periods of time without failure or the need for maintenance. Fan units with a control switch giving a variable performance and direction of air movement are ideal. It is desirable to fit at least two fan units on separate electrical circuits as a back-up facility to allow for the failure or maintenance of one unit. These separate fan units could be of different size and performance thus allowing a wide range of rates of air change (Fig 18). All control switches and connections to the fan units must be waterproof. If the fan unit is installed near a window which can be opened, care should be taken to ensure that the windows are kept closed when the fans are running to prevent short-circuiting of the air circulation with little effect on the general ventilation.

Air inlet(s) should be positioned as low as possible

Figure 18 Extractor fans.
a Electric extractor fan fitted neatly into insulation board. This (and the inlet) should be fitted with an insect-proof screen. Alternatively the extractor fan could have an automatic shutter fitted.

and covered with an insect and rodent-proof mesh screen. If convenient an inlet in the door is suitable. The inlets should be positioned so that noxious fumes are not drawn in. In accommodation within the school building care should be taken to ensure that fresh air is admitted, not air drawn from inside the building. Positioning the inlet in a wall behind a heater can help heat control.

Outlets and extractor fans should be positioned as far away from the inlets and as high as possible. Preferably they should not be in a direct line with the inlets. They could be positioned in a flat roof. The outlet should also be covered with an insect and rodent-proof mesh screen (Fig 18).

Noise

In schools it is unlikely that the sound intensity will seriously disturb the animals. However, two aspects of noise must be considered: the continuous background noise produced, for example, by ventilating fans and general movement, and also that produced by sudden intermittent disturbances. While

Figure 18 b Double extractor fans of different capacity The upper unit is thermostatically controlled.

it is possible for animals to adapt to continuous sound of low intensity with little if any disturbance, sudden sounds may have a serious effect. Everything possible should be done to eliminate or reduce sudden noises and to maintain the background noise level as low as possible.

Exact information about acceptable levels of intensities for animals is not available but, from general experience, it is suggested that the maximum intensity of continuous noise should be 30 decibels, (approximately the intensity of sound in an average living room). The maximum intensity allowed for short periods should be 85 decibels (approximately the intensity of sound produced by a very loud radio).[17, 24]

Light

Three factors related to the provision of lighting equipment should be considered. Firstly, as most researchers note no difference in effect on most animals between artificial light and natural daylight, particularly for small mammals, the provision of windows is open to debate. If it is desired to regulate breeding programmes by controlling the light cycle[16, 21, 25, 36] then the windows will need to be screened, (or some part of the accommodation screened off) to cut out the natural light with its seasonal variation in intensity and duration. Secondly, automatic timing to control the light cycle with a manual overide switch is desirable. One giving an abrupt change on or off of light would appear satisfactory. Thirdly, an intensity control is desirable, but not essential.

It should be recognised that people working in the animal house will have different lighting requirements to the animals.[23] In small accommodation this can be satisfied most easily by the provision of movable lamps to supplement the background illumination. Although an illumination of 105-160 lux (= 10-15 foot candles) is considered sufficient to maintain vital animal activities and rhythms, background mean illumination in a small animal house, measured at the working surface, should not be less than 320 lux (30 foot candles).

With larger houses, where greater human activity will be necessary, the background mean illumination should be between 430-640 lux (40-60 foot candles). Areas where close examination of animals will take place should have a mean illumination of 1076 lux (100 foot candles). As a point of comparison the recommended minimum mean service value of illumination in classrooms is 300 lux and on chalkboards, demonstration benches and in laboratories, 400 lux.[27]

If automatic timing is installed then a light cycle of 12 hours light and 12 hours darkness is generally satisfactory. The light should be provided using incandescent lamps and 'warm white' fluorescent tubes. Light fittings may have suitable diffusers added to reduce glare. All light fittings should be waterproof and fully vapour proof, and readily accessible for easy replacement of the bulb or tube.

Further information on suggested levels of illumination and light period may be found in references 23, 27, 28, 39 and 40.

Humidity

Maintenance of a relative humidity within close limits is only possible with full air conditioning, which is likely to be too expensive to be expected in a school animal house. Humidity control must therefore be relatively crude. Shallow bowls or trays partly filled with water and the residual water after hosing or washing can be used to increase the amount of water in the atmosphere. In turn this can be decreased by increasing the ventilation rate. Alternatively commercially produced electrically operated humidifiers could be used.

The relative humidity should be maintained between 40-60 %.

Humidity affects several of the diseases of some of the animals which are likely to be found in the animal house. Furthermore, it must be remembered that the humidity of the micro-environment in cages and containers will be different from that of the macro-environment. Thus, for example, it is unwise to keep small mammals in containers in which the ventilation is likely to be inadequate such as aquaria and certain types of vivaria, since the humidity may become excessive. It is recognised that rabbits require a lower relative humidity (40-55%) than rats (45-55 %) or guinea-pigs (50-60 %). Thus their cages should be of a more 'open' construction and be placed in a position of low humidity.[28, 35]

Chapter 5 Equipment

Racking and Shelving
Cages with small mammals can be kept on shelves or in racking units, into which the cages slot.

For school use shelving is preferable since it can be cheaper, is simpler and allows the use of a variety of containers. All shelving, and racking, must be capable of carrying loaded cages and containers without excessive distortion. Shelving should be solid, not slatted, to prevent litter material, which is kicked out of the cages by small mammals, from falling into other cages or on to the floor. Large aquaria and particularly heavy cages should be placed on the lower shelves.

Mobile units are useful because they can be moved to facilitate cleaning and to make the layout more flexible. All mobile units should have a secure locking brake (Fig 19).

If at all possible shelving and racking (including their supports) should be made of stainless steel, galvanised iron or aluminium. Although not desirable, wood may be used. However, it must be treated to resist attack by insects and fungi. Cedarwood is a very satisfactory wood to use and requires little, if any, initial treatment. Waterproof plywood could be used for shelves, providing it is protected with a finish of polyurethane varnish or epoxy-resin paint. Do not use cheap timbers that may warp. Timber supports must not touch the floor because they are likely to get wet increasing the chances of rotting.

Metal constructional parts used to make movable shelving should be suitably treated, and then, if necessary, painted and always adequately maintained to prevent rust or corrosion (Fig 20). The ends of hollow tubular parts must be closed.

The back of shelves or racking should be between 7.5 cm and 10 cm (3 inches to 4 inches) away from the wall to facilitate cleaning and allow free air circulation. Mobile units should be positioned so that there is a similar gap between the shelves or racks and the wall. A stop or strip must be attached to the rear edge of the shelf to prevent the cages from projecting and touching the wall. The lowest shelf or rack should be more than 25 cm (10 inches) from the

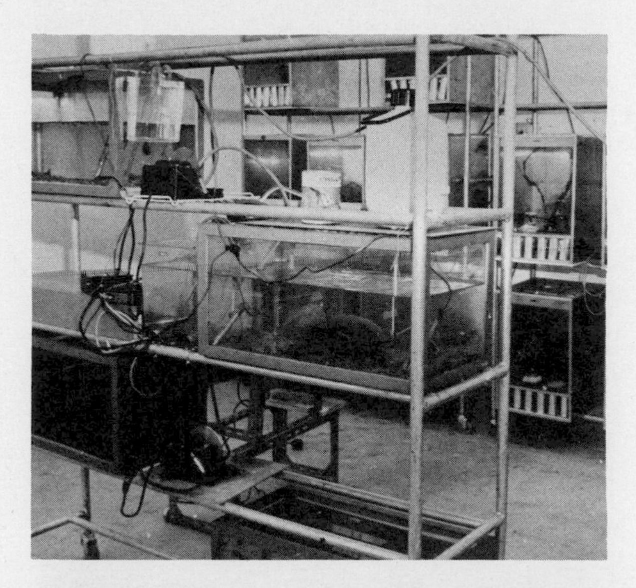

Figure 19 Mobile racking units.
These should be fitted with wheel brakes for safety.

Figure 20 Shelving.
Metal constructional parts for movable shelving must be satisfactorily maintained to prevent corrosion.

Table 1 Approximate spacing required for typical commercial cages

Species or cage	Number of Individuals	Minimum spacing required for one cage in cm		
		Horizontal		Vertical
		Width	Depth	Height
Mouse	Pair	35–40	35	25
Rat/Gerbil Hamster	Pair	50–60	40	35
Guinea-pig	Pair	70–80	40	45
Rabbit	One medium up to 4 kg	80–90	50	60–70
Locust	–	50	50	60–70
Small Finch	One pair	60	40	50
Vivaria	–	60–70	40	45

Note 1. Cages are arranged to fit on minimum shelf depth, most will stand on shelves 30 cm deep. A gap of approximately 10 cm has been allowed for between the cages.

Note 2 2.5 cm approximately 1.0 inch.
30 cm approximately 12.0 inch.

floor and the highest shelf at a height convenient for the smallest user to remove fully loaded cages without using a step ladder.

The length and spacing of shelving or racking can be calculated from the information given in Table 1.

Containers for Organisms

Containers may be broadly grouped into cages or containers suitable for small mammals, for birds, and for insects, vivaria, terraria, aquaria and small glass or plastic jars for protozoan cultures.

Cages suitable for small mammals are of three main types, those having a plastic 'shoebox' base and metal gridding top, those made entirely of metal of mixed sheet and gridding construction and those made almost entirely of wood.

Assessment of the suitability of the design and construction of these cages may be made using the following criteria.

1 Security – against the escape of inmates both adults and young.
2 Safety – for inmates and handlers. There should be no sharp metal edges.
3 Ease of access – to remove and replace inmates.
4 Food and water provision – must be adequate to allow for the longest period between servicing.
5 Ease of cleaning and disinfecting.
6 Size – adequate for the animal(s) housed.
7 Adaptability – it is useful if the cage or its parts can serve more than one function.
8 Labelling – provision of a secure means of labelling on the cage is very useful.

9 Storage – it is convenient if the design allows easy stacking in a minimum space.
10 Economic life – the cage must be safe to use for a long period of time.
11 Visibility – if the inmates are to be observed they must be clearly visible. If the cage is to be used solely for breeding purposes it may be advantageous if visibility is limited to minimise disturbance.

For the smaller mammals such as mice, rats, hamsters and gerbils, cages with plastic bases are most suitable for school use. Guinea-pigs and rabbits may have to be housed in wooden hutches since these are relatively inexpensive for their size. Information on the size of these cages is given in Table 2.

Commercially produced cages are generally to be preferred to home-made ones when considering many of the criteria mentioned above. Cheap cages made of painted thin sheet iron must not be used under any circumstances (Fig 21). An adequate number of spare cages must always be available to facilitate the transfer of animals.

Full details of suitable caging for small mammals will be found in the Educational Use of Living Organisms Project publication 'Small Mammals'.

An excellent summary of the design and construction of cages suitable for small mammals is given in reference 19. Further information may also be found in references 41, 48, 50, 51, 60 and 61.

Cages suitable for small birds such as finches may be constructed from kits or parts which are available commercially.

Containers suitable for insects range from commercially available metal locust cages and butterfly cages to simpler home constructed cages. Information on the construction of such cages may be found in references 46 and 51.

Amphibia and reptiles are usually kept in vivaria or terraria. These are containers in which a terrestrial or semi-aquatic environment is simulated and frequently contain a separate heat and light source, usually an incandescent light bulb. Such containers are available commercially from the major biological supply houses. Information on the construction of vivaria may be found in references 45, 49, 51, 54, 56 and 58.

Aquaria may be constructed either from glass or plastic sheet supported in a metal frame or entirely from plastic. The frames are best constructed of stainless steel or plastic coated metal since they will

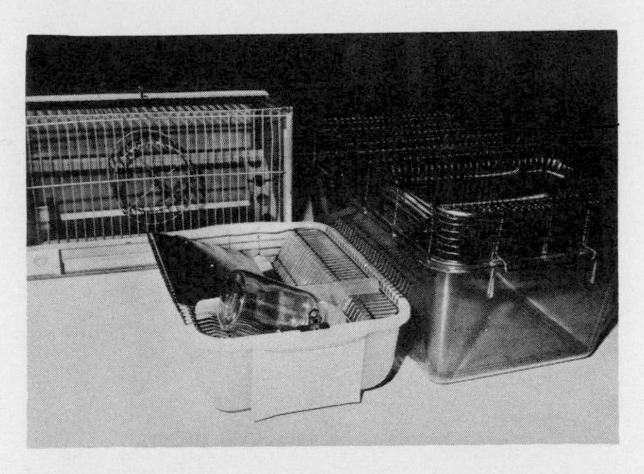

Figure 21 Caging.
A selection of cages for small mammals. The cage in the left background is made of painted sheet iron and is unsuitable for school use.

Table 2 Guide to suggested minimum dimensions for cages in schools

Species	Minimum floor area cm^2	Minimum height cm
Mouse (pair or trio)	500.0	12.5
Rat, gerbil, hamster (pair)	1000.0	20.0–25.0
Guinea-pig (pair)	2500.0	30.0
Rabbit (one up to 4 kg ≏ 9 lb)	5500.0	45.0
Small birds (pair)	3500.0	45.0
Small amphibia/reptiles (two)	3500.0	30.0
Small amphibia/reptiles (one)	1100.0	30.0
Locusts (up to 20 pairs)	1500.0	40.0

Note 2.5 cm approximately 1.0 inch;
6.5 cm^2 approximately 1.0 square inch.

be resistant to corrosion. Plastic aquaria are now available in a range of sizes and are light and convenient in use. However, they are not as easy to clean as those with glass and the surface is likely to become scratched. Aquaria and aquarium frames are available from the major biological supply houses and

Table 3 Suitable positions within the accommodation for particular stock or species

Stock or Species	Position
Protozoan cultures	Some species may require light; place near north facing window or on upper shelves.
Insect cultures excluding locusts.	Away from stored food; on top or bottom shelves.
Locusts; small birds	Place cages near and below extractor fans.
Xenopus	Lower shelves away from door(s) and sink(s) etc.
Other amphibia and reptiles	May require higher temperatures; place on upper shelves.
Small mammals (in general)	Keep in groups of same species; at a height convenient for easy removal of cages.
Rabbits	Require lower temperature and humidity; place cages on lower shelves.
Guinea-pigs	Place cages on lower shelves; may be kept in open pens on the floor if space available.
Breeding stock and genetic stock.	Place in position of minimum disturbance away from door(s) and sink(s).

shops specialising in supplies and equipment for tropical fish.

Designs of a variety of cages suitable for school use may be found in references 46, 47, 51, 55, 63, 64 and 66.

Information on suggested minimum dimensions for cages is to be found in Table 2 and on the positioning of these within the animal house in Table 3.

Work surface

A small work surface or table is needed on which animals can be examined. It should be between 75 cm and 90 cm high to enable the animals to be handled easily and its surface should be covered with a suitable plastic sheeting or hard-wearing washable finish. It could with advantage be mobile. If small mammals are to be examined a sheet of cork or similar material should be available as a rough surface on which these animals can grip.

Small drawers may be fitted under the work surface for the storage of small pieces of equipment.

Trolleys

Trolleys may be easily made using commercially available metal constructional parts, and will be found very useful for transporting cages, food and bedding, etc. Trolleys should be of a size convenient for ease of movement in and out of the accommodation and preferably fitted with a brake or locking device. Metal trolleys must be carefully maintained to prevent rusting and corrosion. The potential hazard of their narrow wheels locking in metal gridding must be recognised and avoided.

Storage

Storage bins of adequate size must be available for food, litter and nesting materials in the animal house. Storage outside the accommodation is not desirable since it is likely to be less secure and materials may become contaminated. Containers for the storage of food, litter and nesting materials must be proof against the entry of insects and rodents and preferably air-tight. Plastic dustbins with twist-lock lids are very satisfactory. It is convenient if the storage bins are on wheeled bases. All storage materials should be kept cool and hygroscopic materials must be kept dry. The storage of large quantities of pelleted food for small mammals is not advisable since this is likely to deteriorate. Stored nesting materials such as hay represent a fire risk.

Soils, composts and chemicals such as fertilisers and pesticides must never be stored with or near the animals' foods, litter or nesting materials.

Cleaning materials and disinfectants

The interior walls of the house including the shelving or racking and containers must be periodically cleaned and disinfected. The shelving should be cleaned and disinfected every fortnight, the rest of the interior of the house at least once a term. Cages and containers, depending on the animals housed, may have to be cleaned and disinfected at least once a week. In general small mammal cages will require cleaning and disinfecting not less than once a week. If disease is present immediate thorough cleaning and disinfection of the entire animal house is recommended.

As a general rule cleaning should always precede disinfection to remove most of the organic matter which tends to inactivate the disinfectant. Soap or liquid detergents are suitable aids to cleaning, powdered detergents are not recommended since they may not dissolve completely and then will persist and contaminate food and bedding. When cleaning, protective clothing, particularly flexible gloves, should be worn. Special soap or liquid hand cream, which may be medicated, should be available. Disposable paper towels are recommended since their use reduces the chance of transmission of infection.

The ideal disinfectant would satisfy the following criteria.[52]

1 It would kill all pathogens likely to infect animals being viricidal, mycoplasmacidal, bactericidal, fungicidal and oocystacidal.
2 It would not be toxic to animals.
3 It would not be dangerous for the operator.
4 It would not be corrosive.
5 It would not be easily inactivated by dirt.
6 It would have a high surface activity and penetration easily removing infected dirt and leaving a protective film when dry.
7 It would not give rise to resistant strains of bacteria, viruses, or fungi.
8 It would be easy to apply.
9 It would be economic in use.

Few disinfectants, if any, will satisfy all these criteria. For school use, chemical disinfection and sterilisation by immersion in suitable solutions is recommended, and therefore an important requisite is that the disinfectant should not be dangerous to the user. For this reason, formaldehyde, mercury salts, phenols and cresols and most halogens, including hypochlorites and bleaches, should not be used. Surface active compounds having a disinfectant and detergent action are among the safest disinfectants and best used. Two types are recommended: quaternary ammonium compounds, e.g. Cetrimide BP and 'Cetavlon' and ampholytic or amphoteric compounds e.g. 'Task', 'Tego' MHG, Griffin 'ASAB', and Harris 'BAS Cleaner'.

Cages can be disinfected by total immersion overnight in a freshly prepared solution of appropriate disinfectant at suitable concentration in the dunk tank or sink.

General information on disinfectants and their mode of action may be found in references 42, 57 and 65. Information on the disinfectants likely to be used in school may be found in reference 86.

Protective clothing

It is advisable to have protective coats or overalls and rubber or plastic gloves available. Head covering and rubber boots may be necessary.

Further information on safety including the use of protective clothing may be found in reference 44.

Fire extinguishers

Fire extinguishers suitable for dealing with electrical fires etc. should be available. Their spray should not be toxic to animals. The dry powder type are suitable.

Incinerators

Disposal, by incineration, of dead bodies and soiled litter and nesting materials etc., is recommended. If it is not possible to dispose of these in the school boilers then the purchase of an appropriate incinerator should be considered.

First Aid Materials

Simple first aid materials suitable for staff and pupils and the animals should be available. A mild disinfectant cream will be suitable for the treatment of minor cuts. Indiscriminate use of proprietary medicines is not recommended.

The box containing the first aid materials should be clearly labelled and placed in an obvious position. Inside the lid of the box there should be full details of the nearest places where medical or veterinary aid may be obtained quickly.

The contents of the first aid box may be similar to those which should be in every laboratory. The contents should include:-

Dressings

Adhesive plaster rolls, 1.25 cm x 5 m
Bandages, 2.5 cm roller
Bandages, 7.5 cm roller
Cottonwool, 15 g packets
Dressing, prepared medium sterile
Gauze, plain white

Triangular bandages.
Wound dressings, assorted adhesive

Equipment
Eye bath
Kidney basin, 15 cm (6 inch)
Medicine glass, graduated
Safety-pins (assorted, rustproof, in box)
Scissors 12.5 cm (5 inch) blunt tipped.

Solutions and chemicals
(All should be clearly labelled)
Acetic acid, 1 % aqueous
Calcium carbonate, powder (dose 15 g stirred up in a ¼ tumbler of water)
Limewater
Sodium bicarbonate
a) 1 % aqueous solution (eye wash) b) 5% aqueous solution
Sodium chloride (dose 2 tablespoonfuls in a tumbler of tepid water).

Nail clippers
The claws of small mammals must not be allowed to grow to a length that they curl underneath the foot. This is most likely with guinea pigs and rabbits. Long claws should be clipped about 5 mm from the quick of the nail in which there are blood vessels.

Weighing scales
Scales of a type giving a direct reading will be found most suitable.

Chapter 6 Health and management

One of the most important considerations in managing a school animal house is the health of both the animals and the people who use them. It needs a balanced approach, neither ignoring that there are potential health hazards, nor the fact that, provided a few simple precautions are taken, these hazards are minimal and that animals can be safely used in schools.

Pupils as well as staff should accept their responsibility for the health of the animals. This can be seen as part of their education, developing a sensitivity to and consideration for the needs of living organisms and an understanding of what is involved in the humane treatment and study of animals.

Human Health

It should be recognised that emotional disturbance as well as hazards to physical health may arise during contact with animals.

Emotional disturbance of pupils as the result of their perception of apparent cruelty or inhumane treatment can be very upsetting. The liability of pupils to be disturbed is very variable depending as it does on their sensitivity to and interpretation of treatment of animals. The possibility of this disturbance is slight with suitable handling and humane treatment of the animals.

Hazards to human physical health are potentially more serious and arise from specific diseases called *zoonoses* transmitted between animals and man, from *allergic reactions* due to the presence of an animal or its products, for example, fur or feather, from *bites* and *scratches* by the animal and from the *handling of equipment and apparatus* required for their maintenance or use.

Zoonoses are due to viruses, mycoplasmas, bacteria, fungi, protozoan animals, parasitic 'worms' or parasitic arthropods. Examples include ornithosis or psittacosis, a virus disease commonly associated with members of the parrot family; salmonellosis caused by bacteria of the *Salmonella* group carried by a variety of animals including tortoises, turtles, cage birds and small mammals; ringworm, a fungus disease

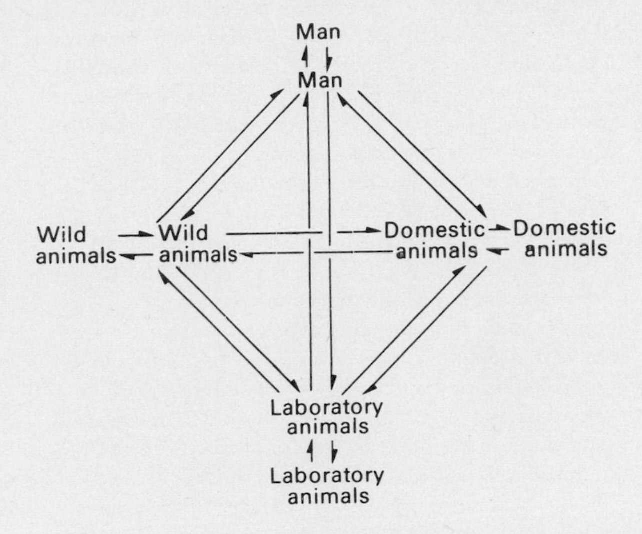

Figure 22 Possible pathways in the transmission of zoonoses.

of the skin carried by a variety of small mammals and mange caused by small mites in the skin again carried by a variety of small mammals.

If an animal is infected with a zoonosis there is a risk that this will be transmitted either to man or to other animals usually of the same species or group. Such transmission to man may be through direct contact for example when an animal bites or urinates on an open wound or when it is handled. Indirect transmission may take place after hand to mouth contact when the hands are contaminated or in the handling of contaminated equipment. Fig 22 illustrates the possible pathways in the transmission.

Further information may be found in references 86 and 128.

Allergic reactions which may occur in certain individuals, are a result of a hypersensitive reaction to certain chemical substances or allergens. These may be produced by animals and other living organisms or may be of non-living origin and include, for example, animal fur, dust from hay or peat, certain detergent washing products and the adhesive in first aid dressings. The onset of sensitivity may be gradual, symptoms may include a reddening and possible swelling and irritation of the skin after dermal contact or an increased flow of secretions in the nose and to the eyes and possible respiratory distress if the allergen is in suspension in the air. Extreme reaction may lead to severe shock. Since the duration of exposure to the allergen is important this should be restricted or prevented with sensitive individuals. Allergic reactions can also be controlled by the use of drugs given under medical supervision.

Further information on allergy may be found in references 90, 92, 97, 104 and 112.

Bites and scratches from animals are most likely when the animal is not tame or is handled incorrectly. Unless the animal is infected such wounds are unlikely to be dangerous. However, the possibility of infection must always be guarded against, all wounds should be cleaned and a suitable dressing applied.

Hazards may also arise in the handling of associated equipment and apparatus. Infection may arise after cuts or abrasions from equipment which is inadequately cleaned or dirty. Cages in particular should not have sharp projections and those which are rusty or badly corroded should be discarded. Possible electrical hazards should be reduced by the careful earthing of all apparatus connected directly to the electric mains.

Animal Health

The animal stock must be kept in the best possible state of health. New animals when obtained should be free from disease and the risk of possible subsequent infection reduced by keeping their surroundings clean and by feeding with uncontaminated food.

The animals must be housed in suitable containers being protected from undue contamination and disturbance and in adequate environmental conditions.

The stock should be subcultured at regular intervals to ensure its vitality.

By keeping animals in a special animal house they are surrounded by a 'barrier' through which passage is controlled to ensure continuance of good health.

Principles of Management

The following principles should be adhered to in order to maintain the best standard of health in the animal stock and to minimise any hazard to human health.

1 Purchase only normal animals which as far as is known are healthy and free from disease. To ensure this they should be obtained from reputable dealers and breeders. Obtain animals from another school only if health of the stock can be assured. Reputable breeders of small mammals co-operate with the Laboratory Animals Centre of the Medical Research Council in an accreditation scheme under which stocks are tested at regular intervals to ensure that they are free from the more important infectious diseases. A list of these breeders may be obtained from the Laboratory Animal Centre and it is advisable that all small mammals are bought from these accredited sources (see Appendix 1, section 1).
2 House all animals in suitable cages of adequate size. Incompatible species should not be housed together.
3 Provide a balanced diet with good quality clean food.
4 Use only uncontaminated litter and nesting materials.
5 Control the macro-environment so that it is kept within the recommended limits.
6 Clean and disinfect all cages and other containers, shelving and/or racking and the inside of the house at appropriate regular intervals (see page 24).
7 Encourage the humane treatment, regular correct handling and sympathetic use of all animals on all occasions.
8 Prevent contact between the stock in the animal house and wild or other domesticated animals. It is particularly important that wild rodents do not gain entry to the house since they may spread disease (see page 27 and Fig 22).
9 Insist on those pupils and staff using the animal house being in a satisfactory state of health and personal hygiene. Do not allow any person to be in contact with the animal stock if infected or with open sores or wounds on the hand. Insist on hands being washed before and after handling animals (see page 15 and Fig 23). Warn people against the dangers of hand to mouth contact when animals are being handled. Use protective clothing.

Teachers, technical staff and pupils who have very frequent contact with the animals would be advised to have a full primary course of injections

Figure 23 Correct hygiene.
"Wash your hands before and after handling animals."

of tetanus toxoid and booster doses after every five years as a precaution, unless they are known to have had such injections within the last five years.

10 Inspect all the animals regularly for departures from normal appearance or patterns of behaviour which may indicate stress or ill-health.

11 Isolate animals suspected of being ill, seek veterinary advice where appropriate if the symptoms are severe or if there are no signs of an improvement in health after, say, two days. Ensure that animals critically ill or distressed are humanely killed by a member of the staff.[84]

If an outbreak of disease is identified by a suitably qualified person then all infected animals must be humanely killed and incinerated along with any infected litter and nesting materials. Cages and shelves should then be cleaned and disinfected.

In serious outbreaks of disease the entire animal stock may have to be destroyed and the accommodation fumigated. The medical practitioners of all staff and pupils who have been in contact with seriously infected animals should be notified in writing.

12 Replace animals at regular intervals to keep a young and healthy stock. Avoid in-breeding unless this is deliberately done for work in genetics.

13 Remove dead animals promptly, together with soiled litter and nesting material and incinerate them. Alternatively the Local Public Health Authority can be notified and arrangements made for disposal. Such materials must not be disposed of in the school refuse containers.

14 Keep a written record of all accidents and illness in the animal house whether they are to the animals or people. Attend to injuries promptly and, if severe, seek medical and/or veterinary advice. Clean the wounds and cover with a dry, clean dressing.

Any unusual symptoms in persons who have been in contact with, or have been bitten by animals, such as fevers or inflammation round the bite or cuts, should be reported to the person's medical practitioner and/or to a hospital if necessary.

15 Allow animals to be taken home by pupils during holiday periods only if absolutely unavoidable and if their health can be safeguarded. Keep these animals separate from the main stock for a few days on their return.

Small mammals should not go to homes where there are other domesticated animals such as cats or dogs in view of the risk of spread of disease. It is equally important to ensure that the feeding and cleaning routine is adhered to.

16 Maintain the animal house and its fittings, services and equipment in good condition. Regularly check all electrical fittings and apparatus.

Chapter 7 Use

If an animal house is constructed and managed appropriately it can serve as a unique and valuable educational facility. It need not just be a place where animals are stored away—albeit healthy and comfortable – except for their visits to the laboratory or classroom.

It is suggested that the purposes of a school animal house should be:

1 To enable a variety of species suitable for school use to be maintained in the best environment obtainable so that they remain in good health.
2 To provide reception and quarantine facilities for newly arrived healthy animals.
3 To regulate contacts between the animals and the people in the school.
4 To provide a supply of normal healthy animals for use within the school.
5 To provide a place in which animals may recover from stresses arising from their use with pupils and where their health can be checked.
6 To provide facilities for pupils to observe animals without unduly disturbing them and where correct animal husbandry can be practised and its related biological principles studied.

As a general rule free access into the animal house should be restricted to supervising staff and a relatively small number of pupils competent in handling the animals and well versed in the essential principles of maintenance. A rule of thumb measure is not to have more than one pupil per 1 m^2 inside the house. Note also that the height of shelving, the positioning of heavy articles etc. will need to be determined in relation to those who will have access to the house. Regularity and familiarity are the keys to success in the management of the house. A consistent routine for handling, feeding and other maintenance activities will be most easy to organise and least disturbing to the organisms who, particularly the small mammals, will consequently become tame and easier to use (see Table 4). If those managing the house become familiar with the condition and behaviour of the animals they will be able to detect and take steps to overcome ill-health rapidly.

To help control the access of people to the house, only two or three keys should be available, one of which should be with the person in charge of fire regulations. Sometimes it is advisable to have the lock on the animal house of a different pattern to those in the rest of the school.

Educational Use

An animal house can be used directly for studies of animal husbandry and ecology. These may be undertaken as class work or as project work by individuals or small groups.

Animal husbandry

This involves not only the techniques of maintaining organisms such as those that have been outlined in previous chapters but also a consideration of the biological principles that underlie these. The aim of these techniques and their associated principles is to allow living organisms to carry out reproduction and other fundamental functions such as feeding, sleeping etc. uninterrupted by disease and other imposed stresses. The relationship of the procedures is shown in diagrammatic form in Figure 24.

Procedures in the normal husbandry of animals are directed towards their maintenance in optimum health. They are correctly caged and kept in a suitably controlled environment which is surrounded by a 'protective shield' intended to minimise or eliminate the introduction or transmission of infection.

Within this 'shield' the environment is kept as clean as possible by the use of hygienic procedures intended to ensure a state of cleanliness. In addition to cleaning and disinfection these procedures also include inspection and disposal.

Contact with wild animals, which may be infected or carry infection is prevented by the 'shield'. Healthy new stock only is allowed through it and unhealthy stock is immediately isolated within a separate 'shield', treated and subsequently returned or disposed of.

Examples of some possible lines of enquiry that could be followed using a variety of animals include:

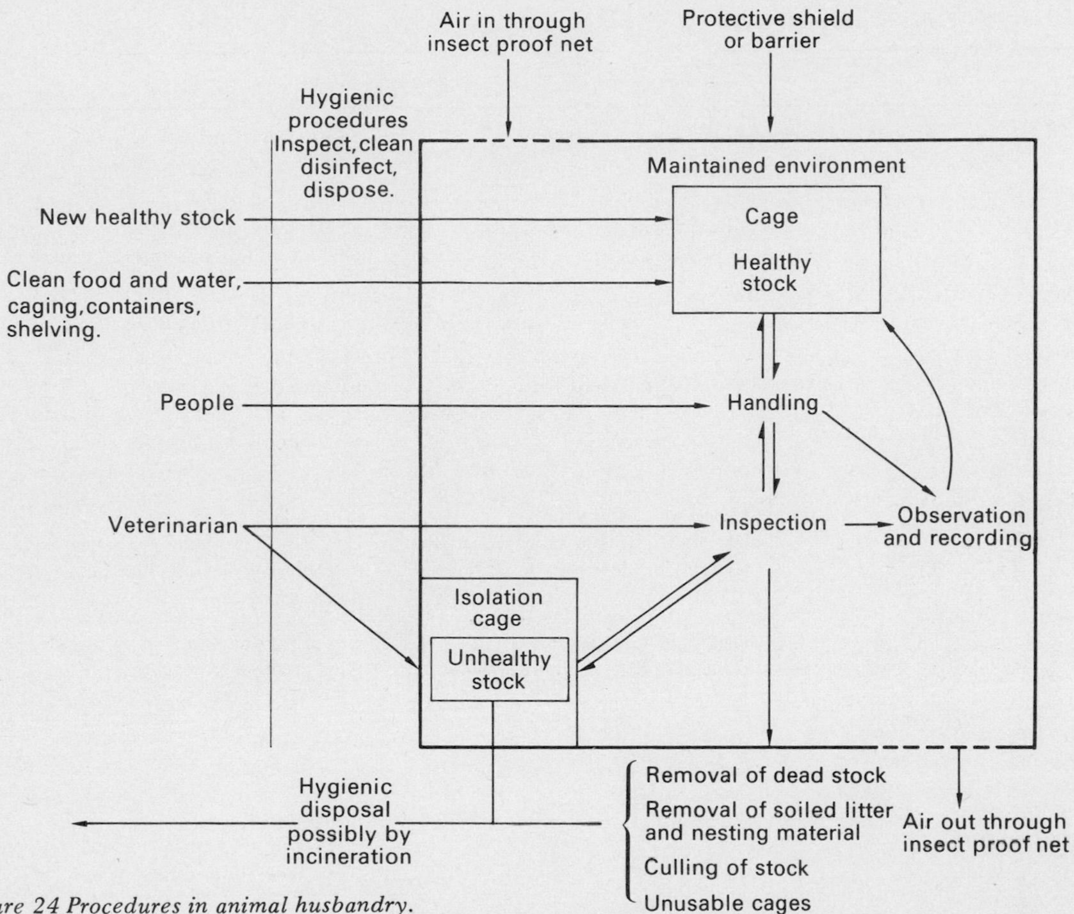

Figure 24 Procedures in animal husbandry.

1 *Maintenance*

What are the essential factors that must be provided for each species to keep them in good health? How can the effectiveness of any particular cage or container be assessed?

How can the person who looks after the animal affect its well-being?

Is a particular routine of maintenance important and, is so, why?

Which foods do the animals prefer?

These questions are possibly best studied by making a pupil entirely responsible for the routine maintenance of a single species. Pupils can then share their experiences with a range of species during discussions in which the teacher may play a guiding role and act as a source of necessary information.[39, 40, 45, 51, 59]

Animals must never be starved or fed on diets deficient in essential components. In any food preference investigation a range of suitable food must be presented. A record can then be made of those eaten and apparently preferred.

2 *Breeding*

What is the natural breeding cycle of a species?

Is this cycle altered by its confinement in a cage or in the animal house?

What changes occur during the period between fertilisation of the female and birth of the young?

What apparently determines the number of young produced by each species? What are the biological consequences of differences in numbers of young produced by different species or different members of the same species?

How are the young looked after following their birth? What is the average rate of growth for the species from birth?

Table 4 Maintenance routine for a typical school animal house

	Monday	Tuesday
Mice	Check water and pellets.	Check water and pellets.
Gerbils	Check water and pellets. Give a small piece of apple or carrot.	Check water and pellets. Give a small piece of apple or carrot.
Hamsters	Check water and pellets. Give a small piece of apple or carrot.	Check water and pellets. Give a small piece of apple or carrot and a few sunflower seeds.
Guinea-pigs Rabbits	Check water and pellets. Give small piece of cabbage, carrot (or apple–guinea-pigs only).	Clean and disinfect cages. Give fresh litter bedding, pellets and water. Give small piece of cabbage, carrot (or apple –guinea-pigs only).
Rats	Clean and disinfect cages. Give fresh litter, bedding, pellets and water.	Check water and pellets. Give small piece of carrot.
Birds	Check seed. Give fresh water.	Check seed. Give fresh water.
Xenopus toads adults		Feed with liver or small fish e.g. guppies. Clean tank and add fresh water.
Fish	Feed once a day with suitable fish food. Check for snails in tanks, take out, crush shells and feed to fish. Clean tanks as necessary.	
Vivarium Lizards Smooth snake Slow worm	Feed with suitable food e.g. insects or slugs. Sprinkle with water.	At least every three months–Clean and disinfect. Give fresh soil.
Locusts	Fresh grass or cereal leaves. If locusts are mating put damp sand into jam jar or metal tube and place in cage, remove when egg pods are evident.	Fresh grass or cereal leaves. At least every month–clean and disinfect cages.
Drosophila @25°C Week ONE	Remove adults from culture at 10.00 Remove males and virgin females and keep separate 16.00	As Monday. Arrange for appropriate matings in labelled tubes.
Drosophila @25°C Week TWO	Remove adults etc. and sub-culture as necessary.	

Wednesday	Thursday	Friday
Check water and pellets. Give small amount of bird seed.	Check water and pellets.	Clean and disinfect cages. Give fresh litter and bedding, pellets and water.
Check water and pellets. Give small piece of apple or carrot and a few sunflower seeds.	1st and 3rd weeks of month. Clean and disinfect cages. Give fresh litter and bedding, pellets and water. Give a small piece of apple or carrot.	Check water and pellets. Give a small piece of apple or carrot.
Clean and disinfect cages. Give fresh litter and bedding, pellets and water. Give a small piece of apple or carrot.	Check water and pellets. Give a small piece of apple or carrot.	Check water and pellets. Give a small piece of apple or carrot.
Check water and pellets. Give small piece of cabbage, carrot (or apple–guinea-pigs only)	Check water and pellets. Give small piece of cabbage, carrot (or apple–guinea-pigs only)	Clean out cage if necessary. Check water and pellets Give small piece of cabbage, carrot (or apple–guinea-pigs only)
Check water and pellets. Give a small piece of carrot and a few sunflower seeds.	Check water and pellets. Give a small piece of carrot.	Check water and pellets, Give a small piece of carrot.
Clean and disinfect cage. Sprinkle fresh sand on base. Give fresh seeds and water.	Check seed. Give fresh water.	Check seed. Give fresh water.
		Feed with liver or small fish e.g. guppies. Clean tank and add fresh water.
Feed once a day with suitable fish food. Check for snails in tanks, take out, crush sheels and feed to fish. Clean tanks as necessary.		
	Feed with suitable food e.g. insects or slugs. Sprinkle with water.	
Fresh grass or cereal leaves. If locusts are mating, put damp sand into jam jar or metal tube and place in cage, remove when egg pods are evident.	Fresh grass or cereal leaves.	Fresh grass or cereal leaves.
Make appropriate matings in labelled tubes.	Check to see if larvae present in mating tubes.	Check to see if larvae present in mating tubes.
	Examine offspring, sex and score.	Examine offspring, sex and score.

	Monday	Tuesday
Tribolium @25°C	Add extra food material to cultures Sub-culture as necessary at least once every three months.	Separate pupae and set up crosses in individual tubes. Examine crosses some 40 days later, sex and score.
Woodlice	Clean container and sub-culture at least once every month.	
Brine Shrimp Pomatoceros Daphnia	Add a small quantity of 'Horlicks' or yeast each day. Do not allow the culture to become cloudy. Sub-culture as necessary.	
Protozoans	Check water level top up if necessary	

Note. The food and water supply for the animals particularly the small mammals should be checked each day to ensure that there is sufficient available and as a matter of priority first thing on Monday mornings.

Weekends: If the animals have to be left unattended during weekends then it is essential that they have sufficient food and, especially, water. Those animals which drink a lot of water e.g. guinea-pigs and rabbits must have a larger supply given on Fridays with plenty of fresh food.

Holiday periods: A basic servicing routine should be kept up ensuring that each animal has sufficient food and water. Stocks should be reduced to a minimum number and breeding should be strictly regulated.

Answers to these questions may be obtained by careful observation, measurement and recording. Relevant sections in the Nuffield 'O' level Biology Teachers' Guides for years II and V may be found helpful.[78, 81] Also see references 39, 40, 51, 59.

3 *Behaviour*

What is 'normal' and what is 'abnormal' behaviour for a particular species in different circumstances?

In what way does behaviour change after cage cleaning, provision of food and water or fresh litter and bedding, or changes in the macro-environment of the house.

Observations of behaviour may be quantified for example by timing certain specified activities e.g. feeding, walking, grooming, 'digging' activity, scratching, examining other occupants in the cage, etc. If such observations are continued over a period of time and careful records kept by pupils then patterns of behaviour can be identified and related to environmental factors.

Pupils should be encouraged to note departures from 'normal' behaviour as they may indicate ill health and should be immediately reported to a teacher or technician.

Further information on the uses of particular groups of animals may be found in the companion publications of the Educational Use of Living Organisms Project.

4 *Hygiene*

What hygienic procedures are adopted in the management of the animal house?

How are they carried out and how effective are they?

What are the possible routes of infection into the animal house and how can they be controlled?

How may infection be transmitted within the animal house?

To assess the cleanliness of particular situations Petri dishes containing sterilised agar can be exposed for a short period of time. Their lids are then replaced and sealed to the base with self-adhesive tape. After incubation the range of types can be identified and the number of colonies can be

Wednesday	Thursday	Friday
		Examine culture. Sprinkle with water. Add fresh carrot or potato.
Add a small quantity of 'Horlicks' or yeast each day. Do not allow the culture to become cloudy. Sub-culture as necessary.		
Sub-culture at least once every three months.		

counted. In view of the possible pathological nature of some of these micro-organisms the lids should not be removed by pupils. When difficulties of visibility arise due to condensation the dish lid should be replaced by the teacher or technician with another dry, sterile lid and then resealed. Discarded cultures should be autoclaved or killed, after separating the Petri dish lid and base, by immersing for a period of time in a freshly prepared solution of appropriate disinfectant at suitable concentration. Alternatively surfaces may be sampled by pressing a strip of self adhesive tape (e.g. 'Sellotape') to them, and then pressing the tape on to the surface of sterilised agar in a Petri dish. These can then be sealed, incubated and examined. Information on suitable bacteriological procedures may be found in reference 78, and in the companion publication *Micro-organisms.*

Ecology

The animal house is a habitat providing both a macro-environment for the organisms and the micro-environments of their cages and containers. Examples of some of the possible lines of enquiry that could be followed include:

1 *Macro-environment*

What is the temperature humidity and light intensity at various points in the house?

What is the air-movement in the house?

How are these factors affected by the distribution of shelving, cages etc. in the house?

Are there seasonal differences?

Humidity may be assessed on a comparative basis by exposing papers impregnated with cobalt salts.[69, 79, 83] Standards for comparison may be prepared by exposing impregnated paper to air of known humidity. Accurate measurement of humidity may be made using wet and dry bulb thermometers. Horizontal air movement may be crudely assessed by suspending a small piece of light wool or cotton thread and noting its angle from the vertical and the direction in which it points.

Vertical air movement may be crudely assessed by using smouldering string as a small smoke generator and noting the direction and relative movement of the smoke. To avoid undue disturbance of the animals the amount of smoke released must be very limited!

Light intensities may be measured using a photographic light meter which will be sufficiently accurate on a comparative basis.

2 *Micro-environment*

In what ways are the micro-environments in each cage or container different from the macro-environment?

How do humans and the animal inhabitants alter the micro-environment?

When does the micro-environment within a cage or container become unsuitable for the animal inhabitant(s) and how does this relate to the procedures of management?

Physical factors in the micro-environment are not easy to measure in many cases without disturbing the environment. Temperature, pH, humidity and

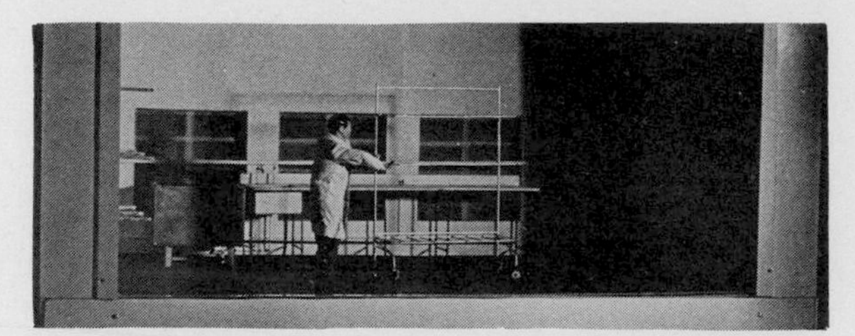

Figure 25 Scale model of animal house. Models like this can be used to plan or study the internal layout.

light intensity may be measured using simple apparatus.[69, 79, 80, 83]

Observation will enable identification of the effect of management procedures on the micro-environment.

3 *Human Use*

In what ways does human use of the house affect the animals and the macro-environment? Do the animals detect the human presence and if so how?

What changes in behaviour occur due to human presence?

Are these changes the same with each person?

What is the best internal layout to facilitate and minimise human movement and how may this be determined?

The movement of people may be studied by making a scale plan board of the house. Small rectangles of wood cut to suitable scale are labelled to represent shelving, sinks, storage bins etc. A panel pin is tapped into these wood blocks so that it protrudes sufficiently for thread to be wrapped round it. The wood blocks are positioned in their correct relative position on the scale plan board. The distance a person travels to carry out simple maintenance procedures can then be measured by attaching a thread to the staring point and tracing subsequent movements by wrapping the thread round the pins in the equipment visited. The length of the thread used gives a measure of the distance moved and an estimate of the energy expended. By altering the arrangement of the scale plan board more effective layouts may be devised. Further information may be found in Nuffield Secondary Science, Theme 6, Field of Study 6.2[82] (Fig 25).

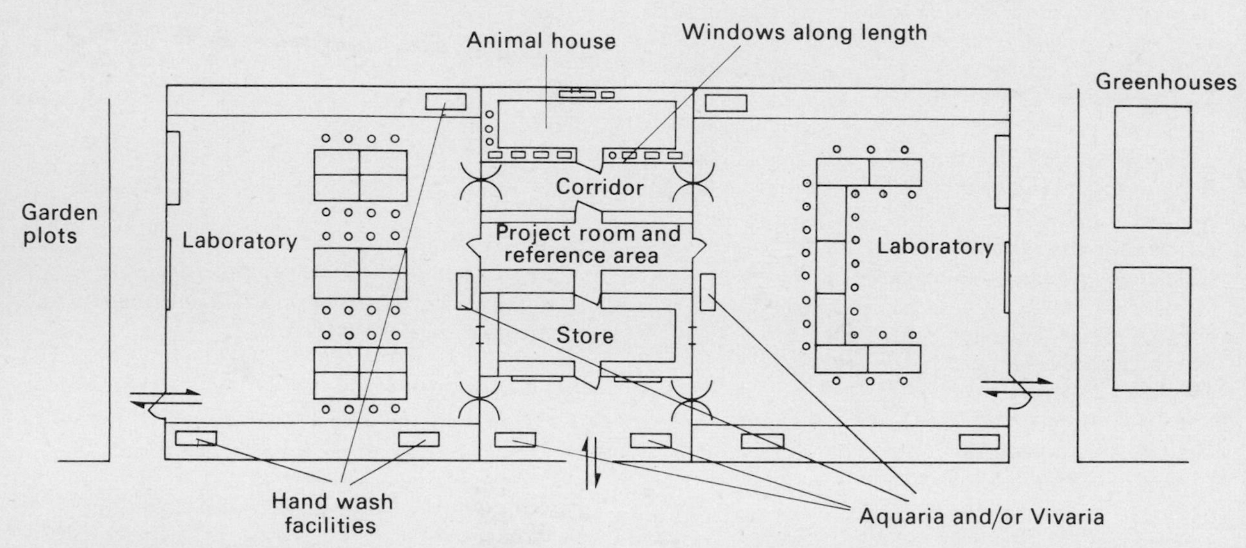

Figure 26 Biology department with integrated animal house.
Integrated animal houses should be centrally situated within the biology department to facilitate transport of stock to and from laboratories and other work areas. In this design observation of the animals may be made from the corridor connecting the two laboratories without interfering with the movement of the pupils in or out of them. Yet the house is remote from the entrance to the department and there need be little student movement along the corridor so that disturbance of the animals will be minimal. Aquaria and/or vivaria are present in the laboratories and in the entrance allowing more room inside the animal house for other species.

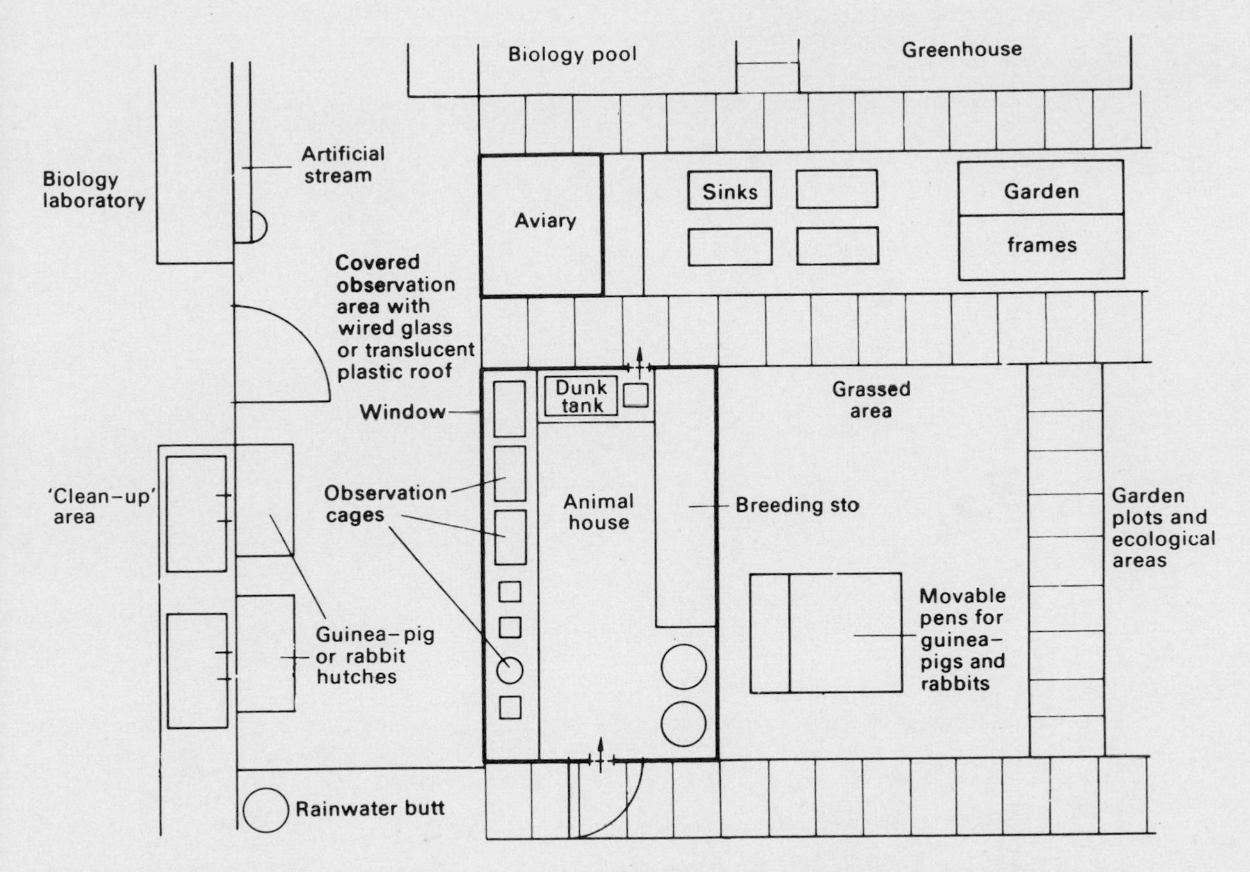

Figure 27 Facility complex adjacent to the biology laboratories.
A detached animal house as part of a facility complex. Observation of the activities of the animals in the house, aviary and hutches may be made from the covered area. If some animals are housed outside in hutches, movable pens and aviary this allows more room inside the animal house for other species.

The Educational Value in the Design and Construction of an Animal House

The building of an animal house by a school can form an excellent project and can involve many departments and pupils.

A suitable design can be considered using the information given in this booklet, specifications can be laid down, appropriate drawings made and then interpreted in suitable building material. Necessary fittings can be installed leaving the connection of the essential services to competent qualified craftsmen.

Prefabricated structures can be easily erected by most schools after the clearance of a suitable site and the laying of the necessary foundations. Suitable insulation can be fixed with the assistance of the handicraft department. Calculation of the required heating capacity can be made, suitable electric heaters installed and connected to the mains supply by a qualified electrician.

To establish the most convenient internal layout scale plan boards and models may be made on which simple work study exercises can be carried out as suggested on page 36. These will enable the best location for storage areas, shelving and other fittings to be accurately worked out (Fig 25).

Educational Use and Design

The use of an animal house for study requires first that the organisms being studied can be separated from those being maintained as stock, for breeding or

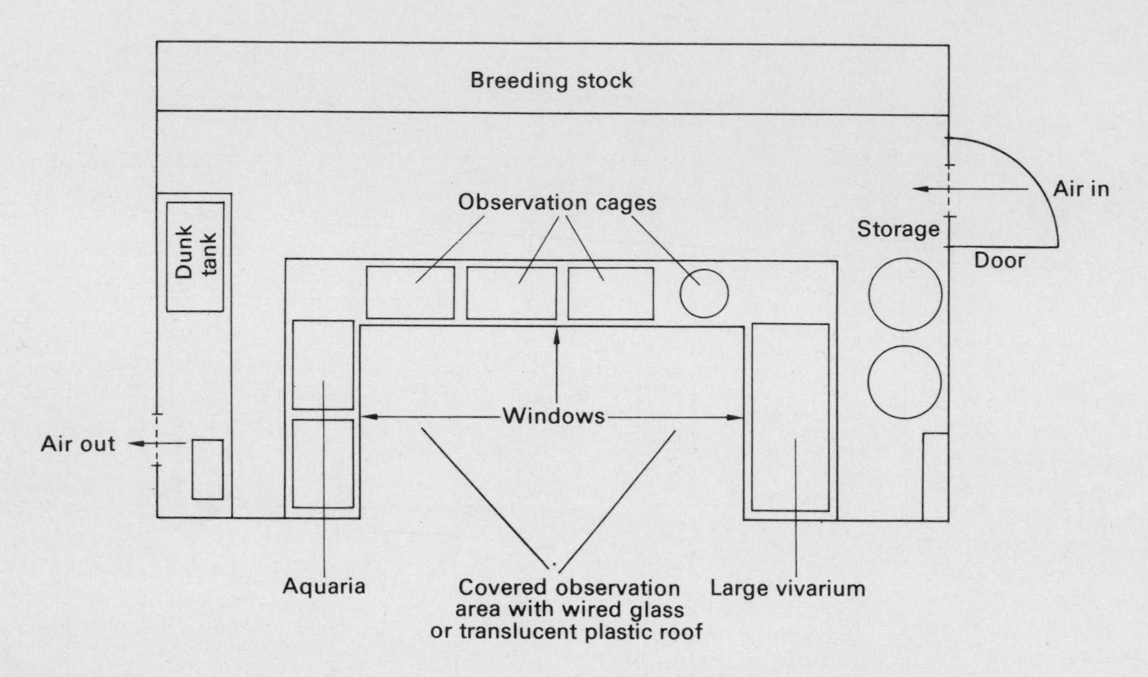

Figure 28 Design for detached animal house.
This design allows observation of the activities of the animals from a covered area with a minimum disturbance to them. Observation can be continued in inclement weather. Detached animal houses should be close to the biology department so that access can be easily regulated and transport of stock to and from the laboratories is not too difficult.

being rested after use, and second that there should be facilities for observing organisms without unduly disturbing them. Three main types of animal house enable these requirements to be met; those built as an integral extension of a laboratory, those which form part of an independent complex of facilities used for keeping and studying organisms, and those which although basically separate maintenance units, have a modified layout enabling some of the organims to be observed from outside. A selection of possible designs are given in Figures 26-28.

Animal houses must be thought of as an essential adjunct to any biology department and positioned so that they do not interfere with laboratory-classroom traffic flow thus allowing pupils the possibility of prolonged observation on the animals.

The final choice of design will be influenced by a number of factors unique to each school situation. Those common factors in design construction and management have been discussed in this booklet.

Appendix 1 Addresses of Sources of Information and Suppliers

1.1 Information

British Standards Institution, British Standards House, 2 Park St., London W1A 2BS.
Building Research Station, Bucknalls Lane, Garston, Nr. Watford, Herts. WD2 7JR.
Centre for Overseas Pest Research (formerly Anti-locust Research Centre), College House, Wrights Lane, London W8 5SJ.
Consortium of Local Education Authorities for the Provision of Science Equipment, (CLEAPSE), Brunel University, Kingston Lane, Uxbridge, Middlesex, UB8 3PH. Information available on, for example, cages for small, mammals, to teachers in Local Education Authorities belonging to the Consortium.
Illuminating Engineering Society, York House, Westminster Bridge Rd., London SE1 7UN.
Institute of Animal Technicians, 16, Beaumont Street, Oxford OX1 2L2.
Institution of Heating and Ventilating Engineers, 49 Cadogan Square, London SW1X 0JB.
MRC Laboratory Animals Centre, Medical Research Council Laboratories, Woodmansterne Road, Carshalton, Surrey SM5 4EF.
School Natural Science Society. Publications obtainable from:- M J Wotton, 44 Claremont Gardens, Upminster, Essex. RM14 1DN.
Scottish Schools Science Equipment Research Centre, SSSERC., 103 Broughton Street, Edinburgh, EH1 3RZ. Information including suggested designs for home-made equipment.
The Under Secretary of State, Home Office H3 Division, Romney House, Marsham Street, London SW1P 3DY.
Universities Federation for Animal Welfare, 230 High Street, Potters Bar, Hertfordshire. EN6 5BP.
Wilson R W (1968) Useful Addresses for Science Teachers, Edward Arnold, London—a compact source of addresses of firms and organisations able to supply information, equipment, materials or specimans.

1.2 Manufacturers/suppliers for wooden sheds and/or concrete garages (including glass fibre cabins).

Banbury Buildings Ltd., Ironstone Works, Banbury, Oxfordshire OX17 3NS.
P.C. Barrett & Son, Portsmouth Road, Ripley, Surrey GU23 6EN.
Ernest Batley Ltd, 193, Sydenham Road, Croydon CR0 2ET.
Compton Associates, Station Works, Fenny Crompton, Nr. Leamington Spa, Warwickshire.
Cotswold Buildings Ltd, Cotswold Works, Standlake, Nr. Witney, Oxfordshire OX8 7QG.
Davies & Clifford Ltd, Star Road, Hillingdon Heath, Middlesex UV10 0UJ
Glass Fibre Developments Ltd, Alpine Works, Newton Road, Crawley, Sussex RH10 2NB. G.F.D. Rediweld Glassfibre cabin.
Robert H. Hall & Co. (Kent) Ltd., 5 Church Road, Paddock Wood, Tonbridge, Kent.
Marley Buildings Ltd, Peasmarsh, Guildford, Surrey GU3 1LS.
Park Lines & Co., 717-719 Seven Sisters Road, London N15 5JU.
Silver Mist Ltd, Reliant Works, Brockham, Betchworth, Surrey RH3 7HT.
D & G Washington, Whiteways, Middle Barton, Oxfordshire.

1.3 Constructional materials and parts for racking, shelving and trolleys

Curties & Baker (Selflok Display) Ltd., Kings Mill Lane, South Nutfield, Redhill, Surrey RH1 5NE. 'Selflok' Aluminium Shelving.
Dexion Ltd., Dexion House, PO Box 7, Empire Way, Wembley, Middlesex HA9 0BR. Metal constructional strip and tube.
Gratnell's Ltd, 256 Church Road, London E10 7JQ. Storage and shelving; trolley frame, racking.
Grundy (Teddington) Ltd., Somerset Works, Somerset Road, Teddington, Middlesex TW11 8TD. Storage bins.
Handy Angle, Link House, 1200, Uxbridge Road, Hayes, Middlesex UB4 8JD. Metal constructional strip.
Metalrax Ltd., Bordesley Green Road, Birmingham B9 4TP. Metal Storage installations.
Remploy Ltd., 415 Edgware Road, London NW2 6LR. Remploy-Lundia shelving.
Savage & Parsons Ltd., Otterspool, Watford By-pass, Watford, Herts. WD2 8HT. 'Spur' adjustable shelving brackets/shelving.
Shelvit, Bellue Vue Mill, Westgate, Burnley, Lancashire. Shelving.
Versatile Fittings, (WHS) Ltd, Bicester Rd, Aylesbury, Bucks. Shelving.
WCB Containers Ltd., Stamford Works, Nayley Street, Stalybridge, Cheshire, SK15 1QQ. Trolleys and bins.

1.4 Manufacturers of cages and racking/shelving units

Ali Type Tools (Woolwich) Ltd., Animal Cages Department, Purland Road, London SE28 0AF.
Aluminium Equipment Co. Ltd., 21a Conewood Street, Highbury, London N5 1B7. Locust cages.
Associated Crates Ltd., Holloway Bank, Wednesbury, Staffordshire WS10 0NL.
E.K. Bowman Ltd., Animal Cages Department, 32, 34 & 57 Holmes Road, London of, NW5 3AG.
Brown Lenox & Co. (London) Ltd., West Ferry Road, Millwall E14 8RY.
Cope & Cope Ltd., Vastern Road, Reading, Berkshire RG1 8BX.
Forth-Tech Services Ltd., Mayfield, Dalkeith, Midlothian. EH22 4AQ.
Hawley D.A., 41a Cottage Grove, Surbiton, Surrey. Bird cages and accessories.
Leicester Aviary Equipment Co. Ltd., 74 Ashfordby St., Leicester LE5 3QG.
National Iron & Wire Ltd., Factory Lane, Blackley, Manchester M9 3AP.
North Kent Plastic Cages Ltd., Home Gardens, Dartford, Kent. DA1 1EQ.
F.W. Potter & Soar Ltd., Beaumont Road, Banbury, Oxfordshire.
Pullman Technical Services Ltd., 75 Tolbooth Wynd, Leith, Edinburgh EH66 EE.

1.5 Manufacturers of food hoppers

North Kent Plastic Cages Ltd., Home Gardens, Dartford, Kent. DA1 1EQ.
S. Young & Sons Ltd., Misterton, Crewkerne Somerset.

1.6 Manufacturers/suppliers of disinfectants and paper towels

Bowater-Scott Corporation Ltd., Knightsbridge, London SW1X 7LR. Paper Towels.
British Hydrological Corporation, Colloidal Works, Deer Park Road, London SW19 3UQ. 'TASK' disinfectant.
British Tissues (Cresco) Ltd., Brent House, 214 Kenton Road, Kenton, Harrow, Middlesex HA3 8BS. Paper towels.
Griffin & George Ltd., Ealing Road, Alperton, Wembley, Middlesex HA0 1HJ. 'Griffin ASAB disinfectant'.
Harris Biological Supplies Ltd., Oldmixon, Weston-super-Mare, Somerset BS24 9BJ. Harris 'BAS Cleaner'.
Hough Hoseason & Co. Ltd., Chapel Street, Manchester M19 3PT. 'TEGO' M.H.G. Disinfectant.
Imperial Chemical Industries Ltd., Pharmaceuticals Division, Macclesfield Works, Hurdsfield Industrial Estate, Macclesfield, Cheshire SK10 2NA. Cetrimide BP, 'Cetavlon'.
Kimberly-Clark Ltd., Larkfield, Maidstone, Kent. ME20 7PS. Paper towels.
Kleenaroll, 13a Church Lane, London N2 8DX. Paper towels.

1.7 Manufacturers/Suppliers of Animal foodstuffs particularly for small mammals

Note that it is likely to be very uneconomic for an individual school to order small quantities of foodstuffs, unless they can collect directly from

the manufacturer or supplier.
E. Dixon & Sons (Ware) Ltd., Crane Mead Mills, Ware, Hertfordshire SG12 9PZ
Fox Facilities Ltd., Home Farm, Aldenham Park, Elstree, Hertfordshire WD6 3AY.
Oxoid Ltd., 20 Southwark Bridge Road, London SE1 9HF.
Pilsbury's Ltd., 21 Priory Road, Edgbaston, Birmingham. B5 7UG.
Spratt's Patent Ltd., Central House, Cambridge Road, Barking, Essex IG11 8NL.

1.8 Suppliers of litter and nesting materials particularly for small mammals.
Note that peat may usually be obtained from local horticultural sundriesman.
F. Clarke & Sons Ltd., 536 Orphanage Yard, Barrington Road, London SW9 7JL
Pilsbury's Ltd., 21 Priory Road, Edgbaston, Birmingham B5 7UG.
Rapnow Ltd., 39 Hartham Road, London N7 9JQ.
W.P. Usher Ltd., 46 Shandos Avenue, Whetstone, London N20 9DX.
Arthur Woollacot Ltd., Blackfriars House, 19-21 New Bridge Street, London EC4 6BD.

1.9 General biological suppliers from whom livestock cages, foodstuffs and litter etc., will usually be available.
Bioserv Ltd., 38 Station Road, Worthing, Sussex. BN11 1JP.
Gerrard & Haigh Ltd., Worthing Road, East Preston, Littlehampton, Sussex. BN16 1AS.
Griffin Biological Laboratories Ltd., Lavendar Hill, Tonbridge, Kent TN9 2BJ.
Harris Biological Supplies Ltd., Oldmixon, Weston-super-Mare, Somerset BS24 9BJ.

1.10 Specialist equipment etc.
Brookwick, Ward & Co. Ltd., 8 Shepherd's Bush Road, London W6 7PQ. Protective clothing, ear punches and tags for small mammals.
C. Ellson & Co. Ltd., UNO Products, Arnold Street, Nantwich, Cheshire CW5 5RB. Immersion Heaters and Thermostats.
The Incinerator Co. Ltd., 14 Coopers Row, Tower Hill, London EC3 N2BJ. Cremators/incinerators.
Imperial Chemical Industries Ltd., Templar House 81/87 High Holborn, London WC1V 6NP. Information on 'Purlboard' insulation board.
Safeburn Ltd., Latchingdon Hall, Latchingdon, Essex CM3 6HD. Incinerators.
Simpsons of Langley Ltd., 88/90 Alpha Street, Slough, Bucks SL1 1OX. Combs, scissors and clippers.
Vent-Axia Ltd., 232/242 Vauxhall Bridge Road, London SW1V 1AU. Fans and ventilating units.
Woods of Colchester Ltd., Tufnell Way, Colchester, Essex. CO4 5AR. Fans and ventilating units.
GEC Xpelair Ltd., P.O. Box 220, Deykin Avenue, Witton, Birmingham. B6 7JH. Fans and ventilating units.

1.11 Trade directories and buyers guides etc.
Sources of information on a variety of manufacturers and suppliers. The main directories are usually available in local libraries.
Advertising section of relevant journals. e.g. Institute of Animal Technicians; Laboratory Animals.
American Association for the Advancement of Science, 1515 Massachusetts Avenue, N.W. Washington, D.C. 20005.
Lists many American suppliers of associated equipment etc. Published annually in November e.g. Guide to Scientific Instruments 1971-1972 Science 174A No. 4010A.
Current British Directories, Ed. Henderson GP & Anderson IG, C.B.D. Research Ltd., Beckenham, Kent BR3 1EA.
Kelly's Manufacturers & Merchants Directory, Kelly's Directories Ltd., Neville House, Eden Street, Kingston upon Thames, Surrey. KT1 1BY. Published annually.
'Where to Buy' Publications, 'Where to Buy' Ltd., John Adam House, London WC2N 6JH.

A more extensive listing of publications and suppliers will be found in the companion publication *The Educational Use of Living Organisms - A Source Book.*

Appendix 2 Estimation of required heating capacity

Estimation of required heating capacity

To keep heating costs down, the energy used to warm the house must be retained by good thermal insulation.

The rate at which heat is lost through different types of building construction is called the thermal transmittance or 'U' value of the construction. The better the insulation the lower will be the 'U' value, as in constructions with high thermal resistance.

Thermal insulation depends mainly on the thermal resistance of the materials used, the internal and external surfaces and any airspaces formed in the construction. The thermal resistance of building materials may be found in references 3 and 29.

To calculate the required heating capacity it is necessary to compute the heat losses from two sources.

1 *Heat loss through walls, roofs and floors etc.*

Rate of heat loss (watts) =

U x area (m^2) x temperature difference (°C) . . . (1)

U for each surface (W/m^2 °C) =

$$\frac{1}{\text{Total resistance to heat loss } (m^2\ ^\circ C/W)} \quad \ldots (2)$$

2 *Heat loss through air changes*

Assuming the air is very rapidly heated:

Rate of heat loss (watts) =

volume (m^3) x number of air changes per second x volumetric specific heat of air (J/m^3 °C) x temperature difference (°C) . . . (3)

Example

A wooden shed is made of deal 25 mm thick with a 13 mm layer of expanded polyurethane insulating material on the walls and roof (excluding window, door and floor, which is standing on brick pillars). There is one window which is 1 m x 0.5 m. The shed is 3 m long, 2 m wide and 2.8 m high and the internal temperature is to be maintained at 20°C, with a minimum outside temperature of 0°C. The volumetric specific heat of air is 1300 J/m^3 °C.

Given the following values for resistances of constructional materials, calculate the required

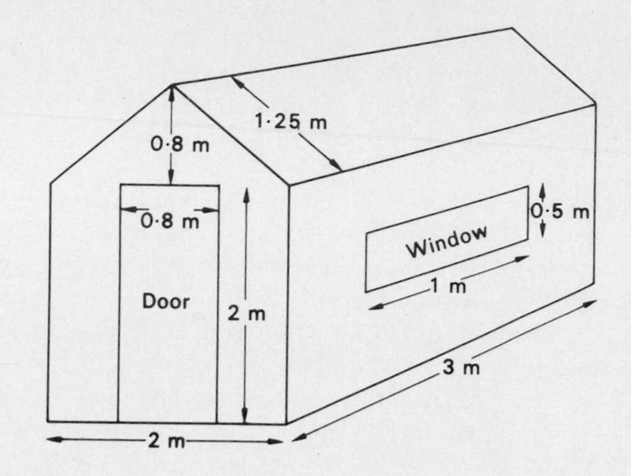

Figure 29 Estimation of required heating capacity. Dimensions of wooden shed.

heating capacity. (Note that these values for resistances are only applicable to this particular case as they take into account the deal and polyurethane thicknesses.)

Resistances:

Wood	=	0.20 m^2 °C/W
Expanded polyurethane	=	0.57 m^2 °C/W
Glass	=	0.003 m^2 °C/W
Floor	=	0.75 m^2 °C/W
Internal surface of walls	=	0.12 m^2 °C/W
" " " glass	=	0.12 m^2 °C/W
" " " roof	=	0.11 m^2 °C/W
External " " walls	=	0.062 m^2 °C/W
" " " glass	=	0.055 m^2 °C/W
" " " roof	=	0.044 m^2 °C/W

Calculating the areas and volumes:

Area:

Floor	3 x 2	=	6 m^2
Roof	3 x 2.5	=	7.5 m^2
Wall (side) ..	3 x 2	=	6 m^2
Wall (side, less glass) . .	6−0.5	=	5.5 m^2
Wall (end) ..	(2x2)+(1x0.8)	=	4.8 m^2
Wall (end, less door) . .	4.8−1.6	=	3.2 m^2
Walls (total area)		=	19.5 m^2

Door	.. 2 x 0.8	=	$1.6\ m^2$
Window	.. 1 x 0.5	=	$0.5\ m^2$

Volume:

Base	.. 2 x 2 x 3	=	$12\ m^3$
Top	.. 1 x 0.8 x 3	=	$2.4\ m^3$
Total volume		=	$14.4\ m^3$

Now the required heating capacity is equal to the overall heat loss, which is made up of the heat loss through the walls, roof and floors etc. plus the heat loss due to ventilation air changes.

1 *Heat loss through walls, roof and floor etc.*
Calculate the U values from equation (2).

(a) Walls (insulated)

Resistance of	internal surface	$0.120\ m^2\ °C/W$
" "	deal	$0.200\ m^2\ °C/W$
" "	expanded polyurethane	$0.570\ m^2\ °C/W$
" "	external surface	$0.062\ m^2\ °C/W$
Total resistance		$0.952\ m^2\ °C/W$

$\therefore\ U = \frac{1}{0.952} = 1.05\ W/m^2\ °C$

(b) Door (uninsulated)

Resistance of	internal surface	$0.120\ m^2\ °C/W$
" "	deal	$0.200\ m^2\ °C/W$
" "	external surface	$0.062\ m^2\ °C/W$
Total resistance		$0.382\ m^2\ °C/W$

$\therefore\ U = \frac{1}{0.382} = 2.62\ W/m^2\ °C$

(c) Roof (insulated)

Resistance of	internal surface	$0.110\ m^2\ °C/W$
" "	deal	$0.200\ m^2\ °C/W$
" "	expanded polyurethane	$0.570\ m^2\ °C/W$
" "	external surface	$0.044\ m^2\ °C/W$
Total resistance		$0.924\ m^2\ °C/W$

$\therefore\ U = \frac{1}{0.924} = 1.08\ W/m^2\ °C$

(d) Floor (uninsulated)

Resistance of	floor	$0.750\ m^2\ °C/W$
" "	deal	$0.200\ m^2\ °C/W$
Total resistance		$0.950\ m^2\ °C/W$

$\therefore\ U = \frac{1}{0.950} = 1.05\ W/m^2\ °C$

(e) Window (uninsulated)

Resistance of	internal surface	$0.120\ m^2\ °C/W$
" "	glass	$0.003\ m^2\ °C/W$
" "	external surface	$0.055\ m^2\ °C/W$
Total resistance		$0.178 m^2\ °C/W$

$\therefore\ U = \frac{1}{0.178} = 5.62\ W/m^2\ °C$

Calculation of heat losses

Using equation (1)

(a) Heat loss from walls	=	1.05x19.5x20.0 = 409.50 watts
(b) " " " door	=	2.62x 1.6x20.0 = 83.84 watts
(c) " " " roof	=	1.08x 7.5x20.0 = 162.00 watts
(d) " " " floor	=	1.05x 6.0x20.0 = 126.00 watts
(e) " " " windows	=	5.62x 0.5x20.0 = 56.20 watts

Total heat loss through structure = 837.54 watts

2 *Heat loss due to ventilation air changes*

Assuming that all the air is very rapidly heated to a uniform temperature of 20°C and that there are six complete air changes per hour (i.e. 6/3600 = 0.0017 air changes per second).

Temperature difference = 20°C − 0°C = 20°C

Using equation (3)

Rate of heat loss = 14.4 x 0.0017 x 1300 x 20
= 636.48 watts

3 *Overall heat loss*

Overall heat loss = heat loss through structure + heat loss through ventilation
= 837.54 + 636.48 watts
= 1474.02 watts

In view of the many additional variables, particularly the degree of exposure of the site and the likely minimum outside temperature, this figure should be taken as an approximation of accuracy ± 20%. It would be wise further to assume that it is on the low side.

$\therefore$ Required heating capacity = 1474.02 + 20%
= 1768.82 watts

Lower outside temperatures will increase the heat loss considerably. In the example quoted the heat loss and the required heating capacity will increase by 5% for every degree Celsius below 0°C.

Appendix 3 Bibliography and References

These are subdivided according to the aspect under which the book or article is referred to or their general nature.

3.1 Structure, Design and Construction.
3.2 The Macro-environment.
3.3 Equipment including Caging and Disinfectants.
3.4 Management and Use.
3.5 Health hazards. Disease.
3.6 Animals and the Law.

The key references are denoted *

3.1 Structure, Design and Construction

1 Architects Journal. Information Sheets (1965) Laboratory floor finishes. *Architect's Journal* 141 (6), 1313-5.

2 Building Research Station. Digest 65 (First Series) *Floor finishes based on polyvinyl chloride and polyvinyl acetate.*

*3 Burberry, P (1970). *Environment and Services.* B T Batsford.

*4 Dowdeswell W H and Kelly P J (1967). The Nuffield Foundation Science Teaching Project–VI: Laboratory Facilties for Biology Teaching. *School Science Review* 49 (167), 5-20.

5 Ferry, D J (1964). *Cost planning of buildings.* Crosby Lockwood.

6 Findlay, G H (1967). Factors considered in designing an animal house, *Journal of the Institute of Animal Technicians,* 18 (1), 29-36.

*7 Hare, R & O'Donoghue, P N (Eds) (1968). *Laboratory Animal Symposia 1. The design and function of laboratory animal houses.* Laboratory Animals Limited, London.

8 Howard, G P (1968). Mechanical Services *in* Hare, R & O'Donoghue P N (Eds) (1968). *Laboratory Animals Symposia 1,* 57-64.

9 Kruk-Schuster, A (1963). Safety aspects in Laboratory design. *Laboratory Practice* 12 (9), 835-7.

*10 Morgan, C G (1968). Electrical Installations in laboratory animal houses *in* Hare, R & O'Donoghue, P N (Eds) (1968) *Laboratory Animal Symposia 1,* 57-64.

11 Sandiford, M (1964). Some problems encountered in the design of an animal house. *Journal of the Animal Technicians Association* 14 (4), 128-33.

12 Statham, S H (1968). Materials for use in laboratory house construction *in* Hare, R & O'Donoghue, P N (Eds) (1968) *Laboratory Animal Symposia 1,* 41-9.

13 Thorp, W T S (1960). The design of animal quarters. *Journal of Medical Education* 35 4-14.

14 Walker, A I T & Stevenson, D E (1967). The cost of building and running laboratory animal units. *Laboratory Animals* 1 (2), 105-9.

*15 Wyatt H V (Ed) (1966). *The Design of Biological Laboratories.* Institute of Biology, 41 Queens Gate, London SW7 5HU.

3.2 The Macro-environment

16 Alexander, D P & Frazer, J F D (1952). The interchangeability of diet and light in rat breeding. *Journal of Physiology* 116 50-51.

*17 Anthony, A (1963). Criteria for acoustics in animal housing. *Laboratory Animal Care* 13 (2), 340-50.

18 Animal Welfare Institute (1958). *Basic care of experimental animals.* Animal Welfare Institute, New York.

*19 Barker, J A (1969). The Environment of small mammals in schools. *School Science Review* 50 (172), 562-6.

20 Bedford, T (1962). Briefing the Architect in Laboratory Animals Centre. *Collected Papers* 11, 65-72.

21 Bruce, H M (1962). Depression in breeding due to short daylight hours *in* Porter G & Lane-Petter W (Eds) (1962) *Notes for breeders of common laboratory animals,* 126.

22 Charles, R T (1953). An experiment to show the effect of strong light on lactating female rats. *Journal of the Animal Technicians Association* 4 (3), 54.

*23 Cobb, K W (1963) Lighting criteria for animal

housing. *Laboratory Animal Care* 13 (3),332-6.

*24 Firman, J E (1966). Heating and ventilation of laboratory animal accommodation. *Journal of the Institute of Animal Technicians* 17 (4), 138-50.

25 Hall, R A (1950). The effect of controlled artificial illumination on rat breeding. *Journal of the Animal Technicians Association* 1 (2), 7-17.

26 Harvey, J W (1968). Air conditioning of animal houses. *Journal of the Institute of Animal Technicians* 19 (1), 1-5.

27 *Illuminating Engineering Society Guide.* (IES Guide) (1970) Illuminating Engineering Society, London.

28 Institute of Laboratory Animal Resources. National Research Council (1968). *Guide for Laboratory Animal Facilities and Care.* U.S. Department of Health, Education and Welfare. Public Health Service Publication No. 1024.

29 Institution of Heating and Ventilating Engineers (1970). *I.H.V.E. Guide.*

*30 Lane-Petter, W (1963). *Animals for research; principles of breeding and management.* Academic Press.

*31 Medical Research Council, Laboratory Animals Centre (1962). The Environment of Laboratory Animals. *Collected Papers* 11. Medical Research Council, Laboratory Animals Centre, Carshalton, Surrey.

32 National Academy of Sciences, Washington D.C., U.S.A. National Research Council and Institute of Laboratory Animal Resources. N.R.C. *Standards for the breeding care and management of laboratory animals. Hamsters* (1960), *Mice* (1962), *Rats* (1962), *Guinea Pigs* (1964).

33 Porter, G & Lane-Petter, W (1962). *Notes for breeders of common laboratory animals.* Academic Press.

34 Porter, G (1965). Some aspects of the physical environment in relation to animal care. *Journal of the Animal Technicians Association* 14 (2), 29-31.

35 Porter, G (1968). Requirements of the animal in Hare R, and O'Donoghue, P N (Eds). *Laboratory Animal Symposia* 1, 15-22.

36 Porter, G, Lane-Petter, W & Horne, N (1963). Effects of strong light on breeding mice. *Journal of the Animal Technicians Association* 14 (3), 117-19.

37 Sainsbury, D W B (1959). *Electricity and problems of animal environment control.* British Electrical Development Association, Publication No. 1897.

38 Scott, P P (1962). How important is the control of environmental temperature in the maintenance of laboratory animals? Laboratory Animals Centre, *Collected Papers* 2, 17-28.

39 Short, D J & Woodnott, D P (Eds) (1969) *The I.A.T. Manual of laboratory animal practice and techniques.* (Institute of Animal Technicians) Crosby Lockwood.

*40 Universities Federation for Animal Welfare 4th Edn. (1972). *Handbook on the care and management of laboratory animals.* E & S Livingstone.

3.3 Equipment including caging and disinfectants

41 Bleby, J (1966). The implications of the rising importance of laboratory animals for the veterinary profession, Laboratory animals—husbandry. *Veterinary Records* 79 (27), 869-73.

42 Breach, M R (1968). *Sterilisation: Methods and Control.* Butterworths.

43 Cohen, B J & Bond E (1952). Cage equipment for laboratory rats and mice. *Proceeding Animal Care Panel 3rd Annual Meetings* 107-17.

*44 Darlow, H M (1967). Safety in the animal house. *Laboratory Animals* 1 (1), 35-42.

45 Hervey, G F & Hems, J (1967). *The Vivarium.* Faber.

46 Hollings M (1967). *Housing Living Things in the Classroom.* School Natural Science Society. Publication no. 26.

47 Jewell, P A (1964). An observation and breeding cage for small mammals. *Proceedings Zoological Society of London* 143 (2), 363.

48 Lane-Petter, W (1952). Mechanics of the animal water bottle. *Nature* 169, 465.

49 Leutscher, A. (1961). *Vivarium Life* 2nd Edn. Cleaver Hume Press Ltd.

50 Mackintosh, J H (1965). The behaviour of small mammals. *Journal of the Animal Technicians Association* 16 (2), 36-8.

*51 Nuffield Foundation Science Teaching Project Junior Science (1967). *Animals and Plants* Ed. M. Hardstaff Collins.

52 Perkins, F T, Darlow, H M & Short, D J (1967). Further experience with Tego as a disinfectant in the animal house. *Journal of the Institute Animal Technicians* **18** (2), 83-92.

53 Sebesteny, A & Williams, P C (1968). A wall-shelving system for rodent cages. *Laboratory Animals* 2 (2), 147-9.

*54 Shaw, P B (1972). *Making Vivaria* 2nd edn. School Natural Science Society, Publication No. 25.

55 Smith, K M (1969). Old balance cases as animal cages. School *Science Review* **51** (174), 102.

56 Spoczynska, J O I (1967). *The Vivarium and the Terrarium.* Nelson.

57 Sykes, G (1965). *Disinfection and Sterilisation.* 2nd Edn. E & F N Spon Limited.

58 Vogel, Z (1966). *Reptiles and amphibians: their care and behaviour* Viking Press, New York.

59 Volrath, J P (1956). *Animals in Schools.* 2nd Edn. University Federation for Animal Welfare.

60 Wallace, M E (1963). Laboratory animals: Cage design, principles, practice and cost. *Laboratory Practice* 12 (4), 354-9.

61 Wallace, M E (1965). The Cambridge Mouse Cage. *Journal of the Animal Technicians Association* **16** (2), 48-52.

62 Wallace, M E (1965). Using mice for teaching genetics. 1 & 2 *School Science Review* **46** (160), 646-658 and **47** (161) 39-52.

63 Welliver, P W (1960). Building animal cages and containers for classroom use. *American Biology Teacher* 22 (8), 489-93.

64 West, R W (1970). A prototype rat cage for use in schools. *School Science Review* **51** (177), 860-3.

65 Whittet, T D, Hugo, W B & Wilkinson G R (1965). *Sterilisation and Disinfection.* Heinemann Medical Books.

66 Wright, D F (1964). The Use of Animals in Teaching. *Science Teaching Techniques* **11** 18-54. John Murray.

3.4 **Management and Use**

67 Animal Welfare Institute (1960) Abuse of animals in the classroom and how it can be avoided. *American Biology Teacher* 22 479-83.

68 Bosse, J J (1962). Live Animals in Teaching. *American Biology Teacher* **24**, 360.

69 Cloudsley-Thompson, J L (1967). *Micro-ecology.* Edward Arnold.

*70 Conalty, M L (1967). *Husbandry of Laboratory Animals.* Proceedings of the Third International Symposium organised by the International Committee on Laboratory Animals. Academic Press.

71 Federation Proceedings (1960). A guide to the literature on the production, care and use of laboratory animals: an annotated bibliography. *Federation Proceedings Federation of American Societies for Experimental Biology* **19** (4), Part III Supplement No. 6 and (1963) **22** (2) Part III Supplement No. 13.

72 Glover, D J & Cotton, D W K (1965). The stability of synthetic vitamin C under various environments. *Journal of the Institute of Animal Technicians* **16** (4), 93-5.

73 Hafez, E S E (Ed) (1970). *Reproduction and breeding techniques for laboratory animals.* Lea and Febiger, Philadelphia.

74 Hastings, J S (1965). The use of vermiculite as a bedding for laboratory animals. *Journal of the Institute of Animal Technicians* **16** (3), 74-6.

75 Hunter-Jones, P (1966). *Rearing and breeding of locusts in the laboratory.* Centre for Overseas Pest Research (formerly the Anti-locust Research Centre), College House, Wrights Lane, London W8 5SJ.

*76 Kelly P J and Wray J D (1971). The Educational Use of Living Organisms. *Journal of Biological Education* 5 (5), 213-8.

*77 Medical Research Council Laboratory Animals Centre (1969). *The accreditation and recognition schemes for suppliers of laboratory animals.* Manual Series No. 1. Medical Research Council, Laboratory Animals Centre, Woodmansterne Lane, Carshalton, Surrey SM5 4EF.

78 Nuffield Foundation Science Teaching Project Biology (1966). *Teachers Guide 2, Life and Living Processes.* Longmans/Penguin.

79 Nuffield Foundation Science Teaching Project Biology (1966). *Teachers Guide 3, The Maintenance of Life.* Longmans/Penguin.

80 Nuffield Foundation Science Teaching Project Biology (1966). *Teachers Guide 4, Living Things*

in Action. Longmans/Penguin.

81 Nuffield Foundation Science Teaching Project Biology (1966). *Teachers Guide 5, The Perpetuation of Life.* Longmans/Penguin.

82 Nuffield Foundation Science Teaching Project Secondary Science. Theme 6. *Movement.* Longmans.

83 Solomon, M E (1945). The use of Cobalt Salts as indicators of humidity and moisture. *Annals of Biology*, 32 75-85.

*84 Universities Federation for Animal Welfare (1968). *Humane Killing of Animals* 2nd edn. Universities Federation for Animal Welfare.

85 Williamson, G R (1971). Animal banking without tears. *School Science Review* **53** (182), 115-6.

3.5 Health Hazards. Disease.

*86 Barker, J A (1969). Hazards of animal maintenance. *School Science Review* **50** (172), 558-62.

87 Bateman, N (1961). Simultaneous eradication of three ectoparasitic species from a colony of laboratory mice. *Nature* **191**, 721-2.

88 Bowden, R S T (1965). The danger of dog and cat worms to children. *Journal of Small Animal Practice* 6 367-70.

89 *British Medical Journal* (1972). Psittacosis. No. 5791, 1st January 1972, 1-2.

90 Centre for Overseas Pest Research (formerly the Anti-Locust Research Centre) (1968). *Note on allergy to Locusts.* Centre for Overseas Pest Research, London.

91 Coombes, R R A & Gell, P G H (1968). Classification of allergic reactions responsible for clinical hypersensitivity and disease *in* Gell P G H and Coombes R R A (1968) *Clinical Aspects of Immunology* 2nd edn., p. 57. Blackwell Scientific Publications, Oxford.

92 Coombes, R R A (1969). The basic allergic reactivity producing disease. *Triangle* (The Sandoz Journal of Medical Science) **9** (2), 43-7.

93 Davies, R R & Shewell, J (1964). Control of mouse ringworm. *Nature* **202** 406-7.

94 Ferris, D H (1960). The biology teacher, animal health and disease in the space age. *American Biology Teacher* 22 (4), 213-20.

*95 Frennes, R (1968). Potentially dangerous pets. *New Scientist* **38** (591), 10-11.

96 Flynn, R J (1967). The control of disease in laboratory animals *in* Conalty, M L (Ed) (1967) *Husbandry of Laboratory Animals.*

97 Frankland, A W (1953). Locust sensitivity. *Annals of Allergy* **11**, 445-453.

98 Gay, W I and Blood, B D (1967). Zoonoses problems and their control in laboratory animal colonies *in* Conalty, M L (Ed) (1967) *Husbandry of Laboratory animals.* Academic Press.

99 Gibson, T E (1967). Parasites of laboratory animals transmissable to man. *Laboratory Animals* **1** (1), 17-24.

100 Halloran, P O (1955). Bibliography of references to disease in wild mammals and birds. *American Journal of Veterinary Research* **10** (2), 1-465.

101 Harris, R J C Ed. (1962). *The problems of laboratory animal disease.* Academic Press. London and New York.

102 Hopkins, H J (1962). Skin disease in laboratory animals caused by fungal parasites. *Journal of the Animal Technicians Association* **13** (2)27-34.

103 Hull, T G (Ed) (1963). *Diseases transmitted from animals to man.* 5th edn. Thomas, Springfield, Illinois, U.S.A.

104 Hunter-Jones, P (1966). Allergy to animals: a zoological hazard. *New Scientist* **31** (513), 615-6.

105 Jenkins, J I & Fletcher, F J (1964). The eradication of lice from a rat colony by means of a malathion dipping routine. *Journal of the Animal Technicians Association* **15** (1), 1-6.

106 Kaplan, C (1970). Rabies. *Science Journal* **6** (4), 32-8.

*107 Loosli, R (1967). Zoonoses in common laboratory animals. *in* Conalty, M L (Ed) (1967). *Husbandry of Laboratory Animals.* Academic Press.

108 MacPherson, L W (1960). Bacterial Infections of animals transmissable to man. *American Journal of Medical Science* **239**, 347-62.

109 Medical Research Council, Laboratory Animals Centre (1961) *Collected Papers.* **10**. Hazards of the Animal House, Medical Research Council, Laboratory Animals Centre, Carshalton, Surrey SM5 4EF.

110 Muul, I (1970). Mammalian ecology and epidemiology of zoonoses. *Science* **170** (3964), 1275-9.

111 Page, K W (1952). The ectoparasites of laboratory and domestic animals. *Journal of the Animal Technicians Association* 3 34-6.

112 Patterson, R (1964). The problem of allergy to laboratory animals. *Laboratory Animal Care* 14 (5), 466.

113 Reichenback-Kilinke, H & Elkan, E (1965). *The principal disease of lower vertebrates.* Academic Press.

114 Schwartz, C (1968). Too much fuss over rabies? *New Scientist* 39 (612), 426-7.

115 Scott, W N (1967). Live Animals in School Teaching. *Journal of Biological Education* 1, 319-23.

116 Scott, W N (1967). Diseases of animals communicable to man. *Biology & Human Affairs* 32 (2), 31-8.

117 Seamer, J (1966). The implications of the using importance of laboratory animals for the veterinary profession, 3 Laboratory animals and disease. *Veterinary Record* 79 (27), 874-8.

118 Sheffield, F W & Beveridge, E (1962). Prophylaxis of 'wet-tail' in hamsters. *Nature* 196, 294.

119 Tomsett, L R (1963). Diseases transmitted to man by dogs and cats. *Practitioner* 191, 630.

120 Tobin, J O'H (1968). Viruses transmissable from laboratory animals to man. *Laboratory Animals* 2 19-28.

121 Twigg, G I; Cuerden, C M & Hughes D M (1968). *Leptospirosis in British Wild Mammals.* Zoological Society, London Symposium 24, 75-98.

122 van der Hoeden, J (Ed) (1964). *Zoonoses.* Elsevier.

123 World Health Organisation (1951). *Joint WHO/FAO Expert Group on Zoonoses: Report of first session.* Technical Report Service No. 40. World Health Organisation, Geneva.

124 World Health Organisation (1959). *Joint WHO/FAO Expert Committee on Zoonoses. Report of second session.* Technical Report Series No. 169, World Health Organisation, Geneva.

125 World Health Organisation (1967). *Joint WHO/FAO Expert Committee on Zoonoses. Report on Third Session.* Technical Report Series No. 378, World Health Organisation, Geneva.

126 Worms, M J (1967). Parasites in newly imported animals. *Journal of the Institute of Animal Technicians* 18 (1), 39-47.

127 Yunker, C E (1964). Infections of laboratory animals potentially dangerous to man: ectoparasites and other arthropods with emphasis on mites. *Laboratory Animal Care,* 14 (5), 455-65.

3.6 Animals and the Law.

*128 Bryant, J J (1967). *Biology teaching in schools involving experiment or demonstration with animals or with pupils.* The Association for Science Education, College Lane, Hatfield, Herts.

129 Council of the Institute of Biology (1966) Observations on the Littlewood Report. *Institute of Biology Journal* (1), 17-20.

*130 *Cruelty to Animals Act* (1876). Her Majesty's Stationery Office, London.

131 Hume, Major C W (1961). The legal control of experiments on animals. *School Science Review* 43 (149), 145-147.

132 Paterson, J S (1966). The implications of the rising importance of laboratory animals for the veterinary profession. I Laboratory animals and the veterinary profession. *Veterinary Record* 79 (27), 864-9.

133 Prosser, L (1960). Code for use of animals in high school biology. *American Biology Teacher,* 22 (8), 478.

134 *Protection of animals Act* (1911). Her Majesty's Stationery Office, London.

135 *Report on the departmental committee on Experiments on Animals* (The Littlewood Report) (1965). Her Majesty's Stationery Office, London, Cmnd. 2641.

*136 Research Defence Society (1969). *Notes on the Law Relating to Experiments on Animals in Great Britain. The Act of 1876.* The Research Defence Society, 11 Chandos Street, London W1M 9DE.

137 Vine, R S (1968). Requirements of the Home Office *in* Hare R & O'Donoghue P N (Eds) (1968) *Laboratory Animal Symposia* 1, 23-7.

138 Williams, P C (1965). The Littlewood Report or the Civil Servants Charter. *Institute of Biology Journal.* 12 (4), 141-5.